STATISTICAL APPROACHES WITH EMPHASIS ON DESIGN OF EXPERIMENTS APPLIED TO CHEMICAL PROCESSES

Edited by **Valter Silva**

Statistical Approaches With Emphasis on Design of Experiments Applied to Chemical Processes
http://dx.doi.org/10.5772/65616
Edited by Valter Silva
Assistant to the Editor: Nuno Tiago Dinis Couto

Contributors

Muhammad Wakil Shahzad, Rahul Davis, Samuel Oliveira, Andre Luis De Castro Peixoto, Rodrigo Fernando Dos Santos Salazar, Ademir Geraldo Cavallari Costalonga, Mateus Nordi Esperança, Ivana Smiciklas, Valter Silva, Muhammad Atiq Ur Rehman, Muhammad Azeem Munawar, Qaisar Nawaz, Muhammad Yousaf Anwar, Alev Yüksel Aydar, Neelesh Kumar Sahu, Atul Andhare, Bamidele Victor Ayodele, Sureena Abdullah

Published by InTech
Janeza Trdine 9, 51000 Rijeka, Croatia

Publishing Process Manager Romina Skomersic
Technical Editor SPi Global
Cover InTech Design team

First published March, 2018
Printed in Croatia
Legal deposit, Croatia: National and University Library in Zagreb

Additional hard copies can be obtained from orders@intechopen.com

Statistical Approaches With Emphasis on Design of Experiments Applied to Chemical Processes, Edited by Valter Silva
p. cm.
Print ISBN 978-953-51-3877-8
Online ISBN 978-953-51-3878-5

For EU product safety concerns:
IN TECH d.o.o., Prolaz Marije Krucifikse Kozulić 3, 51000 Rijeka, Croatia,
info@intechopen.com or visit our website at intechopen.com.

Meet the editor

Valter Silva, BSc, PhD, is presently working as a senior researcher at the Polytechnic Institute of Portalegre, Portugal. He received his PhD degree in Chemical Engineering from the Porto University in 2009. He also obtained an expert's degree in numerical simulation in Engineering by ANSYS and the Polytechnic University of Madrid in 2015. He has published about 60 works in international peer-reviewed journals (a large majority related to advanced numerical and statistical approaches) and conferences. He is the author of seven book chapters and has edited two books. His h-index is 13 with over 450 citations. He is a member of several scientific committees on behalf of the international conferences and several leading international partnerships regarding numerical and optimization projects. He also won several projects and grants in the numerical optimization field with a total funding of about €600K. He has been using his expertise in this field to conduct several successful projects with Portuguese companies.

Contents

Preface

Optimized operating conditions for complex systems can be attained by using advanced combinations of numerical and statistical methodologies. One of the most efficient and straightforward solutions relies on the application of statistical methods with an emphasis on the design of experiments (DoEs). DoE deals with several factors where all of them are varied altogether, instead of one at a time. The great advantage of implementing this strategy is its success to consider multiple interactions between the factors. Furthermore, it also significantly reduces the number of runs necessary to extract meaningful information from data. Since the pioneer work from Box and Wilson where for the first time a systematized approach was developed using DoE to solve optimization problems, this proven methodology was applied successfully to a large number of chemical and other related academic and industrial processes.

Throughout the book, the design and analysis of experiments are conducted involving several approaches, namely, Taguchi, response surface methods, statistical correlations, or even fractional factorial and model-based evolutionary operation designs. This book not only presents a theoretical overview about the different approaches but also contains material that covers the use of the experimental analysis applied to several chemical processes. Some chapters highlight the use of software products to assist experimenters in both the design and analysis stages.

This book helps graduate students, teachers, researchers, and other professionals who are interested in chemical process optimization and also provides a good basis of theoretical knowledge and valuable insights into the technical details of these tools as well as explains common pitfalls to avoid.

The book includes 10 chapters from several researchers and institutions around the world. I would like to express my most sincere gratitude to all the contributor researchers for sharing their work and expertise through this book.

I owe a debt of gratitude to Ms. Romina Skomersic for her outstanding support and help in bringing out the book in the present form.

I'm also indebted to my research team, Mr. Nuno Couto, Mr. João Cardoso, and Ms. Daniela Eusébio for their efforts in all the book stages and valuable suggestions during the review process.

Finally, my special thanks go to the Polytechnic Institute of Portalegre (my institution) and In-Tech-Open Science team for their concern and valuable support to make this book possible.

Valter Silva
Polytechnic Institute of Portalegre

Portugal

Chapter 1

Introductory Chapter: How to Use Design of Experiments Methodology to Get Most from Chemical Processes

Valter Bruno Reis e Silva, Daniela Eusébio and
João Cardoso

Additional information is available at the end of the chapter

http://dx.doi.org/10.5772/intechopen.74061

1. Introduction

Economic pressure and the need to target more competitive levels drive organizations to invest in efficient methodologies to get solutions able to provide clear advantages in a very demanding market. In this scenario, statistical approaches emerge as valuable tools to be used in the chemical process industry. Indeed, the chemical industry uses a wide set of statistical methodologies, ranging from descriptive approaches to complex optimization topics such as Design of Experiments (DoE), always targeting safer, more repeatable and profitable solutions.

Chemical processes often present a complex nonlinear multivariate nature where several factors influence significantly the final outputs. Traditional one-by-one experiment optimization implies the testing of factors one at a time instead of conducting all of them simultaneously. This approach presents several drawbacks, namely requiring an excessive number of experiments, missing the optimal set of factors and neglecting the interactions between the factors [1]. These interactions could play a key role on the system performance. Furthermore, this procedure is very time-consuming. This suggests the DoE approach rather than fundamental or mechanistic models [1–3]. Besides these clear advantages, the DoE implementation is an easy way to reduce the sources of variability in a process as well as is the first step to guide to an optimized solution [3–4].

From a practical standpoint, with DoE implementation, users can find the best solution for any measurable process within corresponding constraints. To do so, the following elements are required:

- An objective function to maximize or minimize a response.
- A predictive model able to describe the main trends of the system.
- Variables then can be adjusted to satisfy the process constraints.

One of the first steps to generate a good predictive statistical model based on DoE is to determine how far it is worth going in the number of factors that really affect the process. Previous data are of major relevance to select a small set of factors. On the other hand, when the users select too many factors, these are often unnecessary and always lead to complex problems to solve.

After selecting the factors and corresponding ranges, the experimental design should be run in a random way. At this stage, empirical models are generated and their adequacy should be evaluated by different statistical procedures such as R measures, analysis of variance (ANOVA) or diagnosis of residual abnormalities [5]. Now, response surface methodology (RSM) should be used to provide the optimal operating conditions for different system responses. This allows generating polynomial functions which determine the minimum, the maximum or a desired value within a range for each response of interest. Optimization can be carried out considering a single response, or taking advantage of the desirability concept, multiple responses with different restrictions [6].

Also, the generated model can be used for robust design purposes [3]. In fact, the computer-aided optimization might set the process on a sharp peak of response and in such manner the system will not be robust to variation transmitted from input variables. Advanced statistical methods (propagation of error is a valuable option among others) can be used to find the flats on response surfaces. These regions are desirable because they do not get affected much by variations in factor settings. Improvements are still possible by narrowing tolerance intervals. To accomplish such goal, several engineering decisions can be taken: (a) accept the response variation as reasonable for this kind of process; (b) change process design specifications; (c) improve the measurement system; (d) improve the control process; or even (e) decline the system as effective to achieve the required targets.

A cost-effective solution could be found narrowing standard deviations from input factors by improving the measurement system or the control process [3, 6]. This is a huge step to become a chemical process more repeatable and predictable, leading to significant money saves.

Previous lines show how statistical models can be used to get optimized and robust solutions in typical problems coming from chemical processes. Additional challenges can also be found and overtaken by using statistical approaches. When the number of significant factors affecting the process is too high, an overwhelming number of runs should be prevented. In such cases, a high fractional design (minimal resolution III or IV designs) could be adopted [4]. To avoid aliases, the minimal resolution designs should be combined with a complete fold-over methodology. This means that a second block of runs with signs reversed on all factors should be included breaking the aliases between main effects and two-factor interactions. Other usual problem found in the chemical industry is to design a set of experiments where both operating and mixture parameters occur. An illustrative example is a baking experiment where besides one observable process variable, six mixture components make part of the input factors [7]. Typical designs do not work at this level. A special type of mathematical formulation is required, and the adequate solution relies on the crossed mixture-process design application. With such approach, it is possible to combine quantitative parameters with mixture component restrictions.

Sometimes, standard RSM designs are not the best option or even suitable to solve chemical problems. Indeed, when multifactor linear constraints or categorical factors are involved, optimal designs are the correct option [6]. This is also the best solution when cubic empirical models are necessary to best fit experimental data.

Many other challenging problems can be faced by using advanced DoE strategies such as nested and split designs, experiments with random factors or even evolutionary operation methods (needed for the continuous improvement of a full-scale process) [2].

Parameter designs (two-array) that are made popular by Taguchi are other suitable option in order to find optimal operating conditions for quality improvement purposes in different chemical processes [8]. Although the approach from Taguchi became extremely popular as an effective tool for quality improvement in 1980s, a large controversy arose because there were significant issues with the advocated experimental strategy and data analysis procedures [9]. Additionally, it was concluded that fractional designs deliver considerable more information becoming much more efficient than the two-array parameter designs developed by Taguchi. Anyway, the use of Taguchi methods is still valid and of utmost importance at industrial level.

Assisting graduate students, teachers, researchers and other professionals by giving them the necessary knowledge on statistical tools with emphasis on DoE approaches that are available to them is perhaps the easiest way to expedite the mainstream of this methodology. In doing so, they can deepen the fundamental theoretical knowledge on the topic as well as optimize chemical processes with more efficient approaches. A more efficient process will be more cost-effective (thus increasing the interest to commercialize it) while improving its performance. Therefore, the aim of this book is to serve as a starting point for new researchers (and experienced ones) wanting to do statistical (emphasis on DoE) analysis of chemical processes.

Author details

Valter Bruno Reis e Silva*, Daniela Eusébio and João Cardoso

*Address all correspondence to: valter.silva@ipportalegre.pt

C3i, Polytechnic Institute of Portalegre, Portugal

References

[1] Anderson M, Whitcomb P. DOE Simplified: Practical Tools for Effective Experimentation. 1st ed. Portland, Oregon: Productivity Press; 2005

[2] Myers R, Montgomery D. Response Surface Methodology. 2nd ed. New York: John Willey and Sons; 2002

[3] Silva V, Couto N, Eusébio D, Rouboa A, Trninic M, Cardoso J, Brito P. Multi-stage optimization in a pilot scale gasification plant. International Journal of Hydrogen Energy. 2017;**42**:23878-23890

[4] Anderson M, Whitcomb P. RSM Simplified – Optimizing Processes Using Response Surface Methods for Design of Experiments. 1st ed. Productivity Press; 2005

[5] Silva V, Rouboa A. Combining a 2-D multiphase CFD model with a response surface methodology to optimize the gasification of Portuguese biomasses. Energy Conversion and Management. 2015;**99**:28

[6] Silva V, Rouboa A. Optimizing the gasification operating conditions of forest residues by coupling a two-stage equilibrium model with a response surface methodology. Fuel Processing Technology. 2014;**122**:163

[7] Ketelaere B, Goos P, Brijs K. Prespecified factor level combinations in the optimal design of mixture-process variable experiments. Food Quality and Preference. 2011;**22**:661

[8] Apparao K, Birru A. Optimization of die casting process based on Taguchi approach. Materials Today: Proceedings. 2017;**4**:1852

[9] Available from: http://citeseerx.ist.psu.edu/viewdoc/download?doi=10.1.1.595.175&rep=rep1&type=pdf [Acessed February 2, 2018]

Chapter 2

Design of Experiments Applied to Industrial Process

Neelesh Kumar Sahu and Atul Andhare

Additional information is available at the end of the chapter

http://dx.doi.org/10.5772/intechopen.73558

Abstract

Response optimization and exploration are the challenging task in front of experimenter. The cause and effect of input variables on the responses can be found out after doing experiments in proper sequence. Generally relationship between response of interest y and predictor variables x_1, x_2, x_3, … x_k is formed after carefully designing of experimentation. For examples y might be biodiesel production from crude 'Mahua' and x_1, x_2 and x_3 might be reaction temperature, reaction time and the catalyst feed rate in the process. In the present book chapter, design of experiment is discussed based on predictor variables for conducting experiments with the aim of building relationship between response and variables. Subsequently a case study is also discussed for demonstration of design of experiments for predicting surface roughness in the machining of titanium alloys based on response surface methodology.

Keywords: design of experiments, response surface methodology, optimization, ANOVA

1. Introduction

Researchers found the unknown solutions by conducting experiments with the help of varying two or more inputs factors [1]. Typical solutions are obtained from experiments are:

- Effect of input variables over the solutions or responses
- Which combination of input variables will give best solution?
- What are ranges of variables suitable for experiments?
- Under what condition should we operate our plant?

Experiments help us to direct compare among treatments of interest. Design of experiments minimizes bias in the comparison which helps in reducing error [2]. One of the advantages in design of experiments that we can control the experiments which allows us to make decision about influence of input variables over the response. Explicitly, one can make conclusion about causation.

An experiment consists of treatments, experimental unit, responses and a method to assign treatments to unit. Mosteller and Tukey [3] describes three concepts for the development of relationship between variables and responses namely consistency responsiveness and mechanisms. Proper design of experiments should avoid systematic error, should be precise, allows estimation of errors and have broad validity.

Some important terms and concepts used in design of experiments are listed below

1.1. Treatment

It defines as are the diverse actions for equate. Amount of fertilizers in agronomy, different long distance rate structure in marketing or different temperatures in reactor vessel in chemical engineering are examples of treatments.

1.2. Experimental units

These are units in which treatments are applied. Graph are plotted for to see variation of these units over response.

1.3. Responses

These are the outputs we measures during experiments. These responses define the mechanism of the process during experiments. Responses for examples might be fatty acid ethyl ester nitrogen content in biodiesel production or combustion performance biodiesel biomass of corn plants, profit by production, or yield and quality of the product per ton of raw material.

1.4. Randomization

It is distribution of variables within the range with recognized, defined probabilistic mechanism for the assignment of treatments to units.

1.5. Experimental error

It is defined as variation present in all experimentally measured responses. Experiments runs on different range of variables will give different results for responses. Moreover conducting experiments at the same range of variables over and over again will give different results in different trials. It should be noted that experimental errors within acceptable range does not indicate conducting wrong experiments.

1.6. Measurement units

It is the unit of measured responses for example combustion pressure in different % blend of biodiesel. These may differ from the experimental units. For example Fertilizer is applied to a plot of land containing corn plants, some of which will be harvested and measured. The plot is the experimental unit and the plants are the measurement units. Ingots of steel are given different heat treatments, and each ingot is punched in four locations to measure its hardness. Ingots are the experimental units and locations on the ingot are measurement units.

2. Design of experiments

An experiment can be defined as a test or series of runs in which purposeful changes are made to the input variables of a system or process so that changes in the output response variable may be observed and the reasons for the same may be identified [4–6]. Some process variables $x_1, x_2, \ldots x_p$ are controllable, whereas other variables $z_1, z_2, \ldots z_q$ may be uncontrollable. An experiment serves the following purposes:

a. Determine which variables $x_1, x_2, \ldots x_p$ are most influential on response y.

b. Determine where to set the influential x's so that y is always near to the desired nominal value.

c. Determine where to set the influential x's so that variability in y is minimized.

d. Determine where to set the influential x's so that effects of uncontrolled variables are minimized.

Design of Experiments refers to the process of planning, designing and analyzing the experiment so that valid and objective conclusions can be drawn effectively and efficiently [7]. In order to draw statistically sound conclusions from the experiment, it is necessary to integrate simple and powerful statistical methods into the experimental design methodology [8]. The success of any industrially designed experiment depends on sound planning, appropriate choice of design and statistical analysis of data and teamwork skills.

2.1. Approaches for experimentation

The approach to planning and conducting the experiment is called the strategy of experimentation [9]. The best guess approach is the most common and uses guesswork to arbitrarily select a combination of input factors for testing. However, this is unscientific and one cannot confirm whether a better response obtained is indeed the best solution.

Another approach is the 'one factor at a time' (OFAT) in which one factor is sequentially varied at a time by different levels and all other factors are kept constant. The levels may be quantitative (such as temperature or voltage) or qualitative (such as presence of coolant). The main effect of the factor is the change in response produced by a change in the level of the factor. However, OFAT approach can show only one causal effect and many a times, the causal effect of multiple factors is not additive, meaning there is interaction between them. An interaction is the failure of one factor to produce the same effect on the response at different levels of another factor. OFAT approach cannot give interaction effects as all other factors are kept constant when a factor is varied.

The scientific approach therefore is to vary several factors together at a time so that both main effects as well as interaction effects of factors on the response variable may be identified and studied. This is called factorial experimental design and this is the only way to discover interactions between variables. In factorial experiments, factors contain discrete values (levels), and the number of factor levels influences design of experimental runs. When all possible combinations of the levels of the factors are investigated, then it is called a full factorial experiment. In contrast, a fractional factorial experiment is a variation of the full factorial design in which only a subset of the runs is used.

Various other kinds of experimental designs are in place such as Plackett-Burman design, Taguchi method, response surface methodology, mixed response design and Latin hypercube design [10]. Each of these designs uses different techniques to generate experimental runs. Of these, response surface methodology is of particular interest as it takes three levels of factors to generate an experimental design sequence and uses a quadratic polynomial model for conducting analysis.

The three principles of experimental design such as randomization, replication and blocking are used in industrial experiments in order to improve the efficiency of experimentation. Randomization is the random ordering of experiments to ensure all levels of a factor have equal chance of being affected by noise factors (unwanted sources of variability) such as temperature or power fluctuation. Replication is the process of repeating all or a part of experiment runs in a random sequence to allow more precise estimation of experimental error as well as main and interaction effects. Blocking is the process of arranging similar experimental runs into blocks (or groups) to distribute the effect of change in blocking factors such as batch, machine, time of day, etc. across the experiments and avoid confounding (confusion whether the output change is due to change in block or change in factor level).

For statistical analysis under design of experiments (DOE), the factor level numbers are considered instead of the actual value of the factor at that level. In other words, the factors are represented by coded variables instead of natural or uncoded variables. In case of categorical variables, the levels are represented in natural numbers as 1, 2, ... l. Quantitative variables can also expressed in this manner in many experimental design methods.

Let x_i and w_i be the coded and uncoded values respectively for a level i of a control variable having l_i levels. Then w_{low} and w_{high} refer to the uncoded values of the factor at the lowermost and uppermost levels respectively. For categorical variables, x_i and w_i are expressed as Eqs. (1) and (2).

$$x_i = \frac{w_i}{(w_{\text{high}} - w_{\text{low}})/(l_i - 1)} \tag{1}$$

and

$$w_i = x_i \frac{(w_{\text{high}} - w_{\text{low}})}{2} \tag{2}$$

In case of response surface methodology, the number of levels for all quantitative variables is odd, and the middle level is given the value 0. Thus the remaining levels get equally distributed on both sides of the middle level, for example, −2, −1, 0, +1, +2. Then, x_i and w_i would be expressed as Eqs. (3) and (4).

$$x_i = \frac{w_i - (w_{\text{high}} + w_{\text{low}})/2}{(w_{\text{high}} - w_{\text{low}})/2} \tag{3}$$

and

$$w_i = \frac{(w_{\text{high}} + w_{\text{low}})}{2} + x_i \frac{(w_{\text{high}} - w_{\text{low}})}{2} \tag{4}$$

3. Response surface methodology

Response surface methodology or RSM is a collection of mathematical and statistical techniques used for the modeling and analysis of problems in which a response of interest is influenced by several variables and the objective is to optimize the response. The method was introduced by G. E. P. Box and K. B. Wilson in 1951. It uses a sequence of designed experiments to obtain an optimal response and uses a second-degree polynomial model to achieve this.

Let a process contain n input variables x_1, x_2..., x_n. Then the response y is given by Eq. (5)

$$y = f(x_1, x_2, \ldots, x_n) + \varepsilon \tag{5}$$

Where, ε is the error or noise observed in the response. If the expected response is denoted by $E(y) = f(x_1, x_2, \ldots, x_n) = \eta$, then the response surface is represented by Eq. (6)

$$\eta = f(x_1, x_2, \ldots, x_n) \tag{6}$$

The response can be represented graphically, either in the three-dimensional space or as contour plots that help visualize the shape of the response surface. Contours are curves of constant response drawn in the x_i, x_j plane keeping all other variables fixed. Each contour corresponds to a particular height of the response surface. RSM also explores relationships the response variables and several input variables. If the response is modeled by a linear function of the independent variables, then the approximating function is the following linear model shown by Eq. (7).

$$y = \beta_0 + \beta_1 x_1 + \beta_2 x_2 + \ldots + \beta_n x_n + \varepsilon \tag{7}$$

If there is curvature in the system, then a polynomial of higher degree must be used. Most of the industrial problems can be modeled with sufficient accuracy by using a second-degree polynomial, which yields the following second order model shown by Eq. (8)

$$y = \beta_0 + \sum_{i=1}^{n} \beta_i x_i + \sum_{i=1}^{n} \beta_{ii} x_i^2 + \sum_{i=1}^{n-1} \sum_{j=i+1}^{n} \beta_{ij} x_i x_j + \varepsilon \tag{8}$$

The method of least square chooses β's in Eq. (8) so that the sum of the squares of the errors ε, are minimized. The least squares function is shown by Eq. (9)

$$L = \sum_{i=1}^{n} \varepsilon_i^2 \tag{9}$$

By putting value of ε_i from above equation and differentiating equation with respect to coefficient β, regression coefficient can be obtained.

3.1. Response surface designs

Response surface designs are those experimental designs which are used for fitting response surfaces and generally contain three factor levels [11]. Two types of response surface designs are used namely, central composite design and Box-Behnken design.

3.1.1. Central composite design

This consists of a factorial design (the corners of a cube), center and axial (or star) points that allow for estimation of second-order effects [12]. The addition of axial points practically increases the number of levels to five as shown in **Figure 1**. This may create problems if the axial points cannot be run due to technical or safety reasons. For a design having k factors, the distance of the axial point from the design center is $\alpha = 2^{k/4}$.

A central composite design containing axial points with the calculated value α is called circumscribed central composite design. If it is not possible to use this value of α, then a provision exists in which α can be taken equal to 1 in order to obtain what is called as face centered central composite design.

3.1.2. Box-Behnken design

This design overcomes some loopholes of central composite design by avoiding axial points and corner points of the design space (or bypassing extreme factor combinations) and by taking only

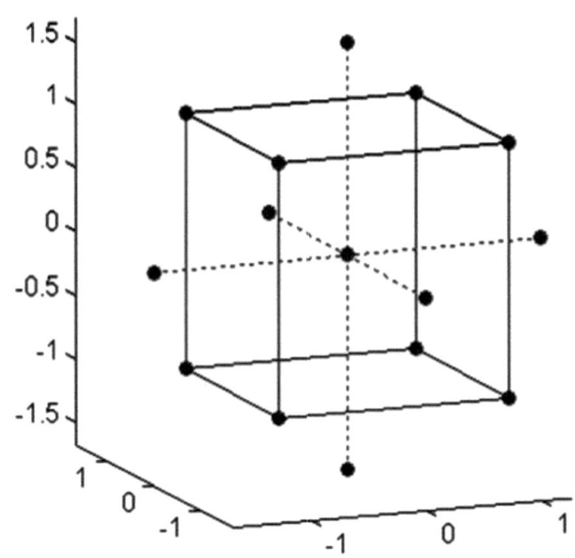

Figure 1. Central composite design for three factors.

three factor levels as shown in **Figure 2**. The design ensures that all factors are never set to high levels simultaneously and thus ensures design points within safe operating limits.

Also, this design is fully rotatable, meaning that it provides the desirable property of constant prediction variance at all points that are equidistant from the design center. Compared to central composite design, this design gives lesser number of experiment runs for the same number of factors. Hence, it can be seen that Box-Behnken designs have several advantages over central composite designs.

3.2. Analysis of variance (ANOVA)

The analysis of variance (ANOVA) established by Ronald Fisher in 1918, is a statistical tool used to analyze variation among and between groups. ANOVA is used to see the significant and insignificant parameters of the predicted model. This procedure involves checking individually variability of variable over the response [13]. It is based on the concept of two hypotheses namely H_0 (means all the regressions coefficients are zero) and H_1 (mean at least one of the regression coefficient is non-zero). If H_0 is false then it suggests that one or more of the variable contribute significantly to the developed model for response [14]. In this test procedure, sums of square of regression and errors are calculated. To verify hypothesis F value is calculated as ratio of mean of square (regression) to mean of square (error) is calculated. Larger values of F suggest that model is significant. Alternatively, p value is the probability of the predicted model shows its significance in terms of statistics. If p value is less than 0.05 model terms are significant and p value greater than 0.05 indicates that model terms are not significant. Similarly the value of R^2 (correlation coefficient) is calculated as ratio of sum of square of regression to the total sum of square. The correlation coefficient (R^2) value suggests a satisfactory representation of process by model and good correlation between experimental and theoretical values provided by the model equation. For goodness of fit of the model, R^2 (correlation coefficient) should be at least 0.80. However, a large value of R^2 does not necessarily imply that the regression model is good one. Adding a variable to the model will always

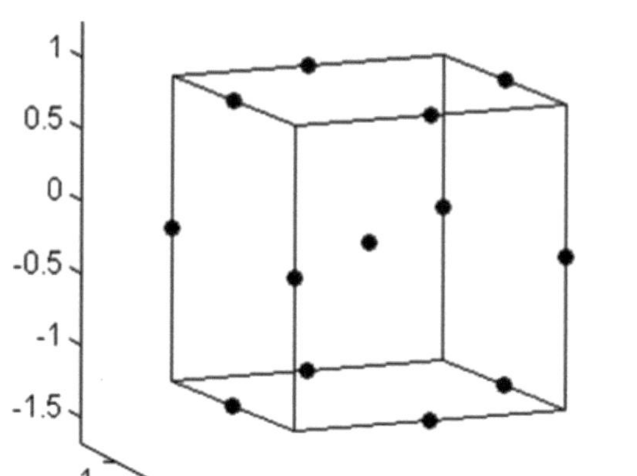

Figure 2. Box-Behnken design for three factors.

increase R^2, regardless of whether the additional variable is statistically significant or not. Thus it is possible for models that have large values of R^2 to yield poor predictions of new observations or estimates of the mean response. Therefore sometimes it is beneficial to calculate adjusted correlation coefficient (R^2_{adj}) which is calculated as (1 – sum of square (error)/sum of square (total)). Once R^2 and R^2_{adj} are different affectedly, there is a decent probability that non-significant terms have been included in the model.

3.3. Backward elimination approach for developed model evaluation

After developing a model, its adequacy is checked by F test and p value [15]. For a model term to be significant it should have high F value and low p value. Insignificant model terms do not affect the response therefore can be removed from the model. In order to avoid insignificant terms in the model such that modified model clarifies the response, the backward regression elimination method (also known as stepwise deletion) is used. In the stepwise deletion method, t test or F test for significance of design variable is performed with sequence begin with full model. Insignificant variables with the highest p value (e.g. $p > 0.05$) are removed from the full model. Stepwise regression procedure details are as follow:

Step 1:

Initially the model can be written as shown in Eq. (10)

$$y = \beta_0 + \beta_1 x_1 + \ldots\ldots. + \beta_{n-1} x_{n-1} + \varepsilon. \tag{10}$$

Then, the following n−1 tests are carried out, for null hypothesis H_{oj}: $\beta_j = 0$. The lowest partial F-test value F_l corresponding to H_{oj}: $\beta_j = 0$ or t-test value t_l is compared with the preselected significance values F_0 and t_0. One of two possible steps (step 2a and step 2b) can be taken.

Step 2a:

For eliminating any variable say x_l, it should satisfy the following case $F_l < F_0$ or $t_l < t_0$. Now the modified model can be written as equation

$$y = \beta_0 + \beta_1 X_1 + \cdots + \beta_{l-1} X_{l-1} + \beta_{l+1} X_{l+1} + \cdots + \beta_{n-1} X_{n-1} + \varepsilon \tag{11}$$

Step 2b:

If $F_l > F_0$ or $t_l > t_0$, the original model is the model we should choose.

The procedure will automatically stop when no variable in the new original model can be removed and all the next best candidate cannot be retained in the new original model. Then, the new original model is our selected model.

In the present thesis, measured responses after machining are analyzed using responses surface methodology with cutting parameters as input variables. Initially RSM models are developed for each response. Significance of each variable is confirmed through ANOVA analysis then insignificant terms are removed using backward elimination approach. Analysis of machining responses is discussed in above sections.

4. Case study for using design of experiments in machining operation

Surface roughness is most widely used indicator to quantify surface integrity of machined part [16, 17]. It directly gives quality of surface finish and has been used by many researchers. Surface roughness is influenced by several factors such as - cutting speed, feed, depth of cut, tool geometry, tool wear, etc. [17–20]. Therefore in the present work surface roughness is taken as response.

In the present case study, design of experiments with central composite design was performed based on response surface methodology. This is constructed as factorial design (the corners of a cube), center and axial (or star) points that allow for estimation of second-order effects [21]. The addition of axial points practically increases the number of levels to five. This may create problems if the axial points cannot be run due to technical or safety reasons. For a design having k factors, the distance of the axial point from the design center is $\alpha = 2^{k/4}$ as shown in **Figure 3**. If it is not possible to use this value of α, then a provision exists in which α can be taken equal to 1 in order to obtain what is called as face cantered central composite design. In the present case study, based on input factors and their levels as shown in, 20 set of experiments were performed, each for turning and milling operations. The design of experiments

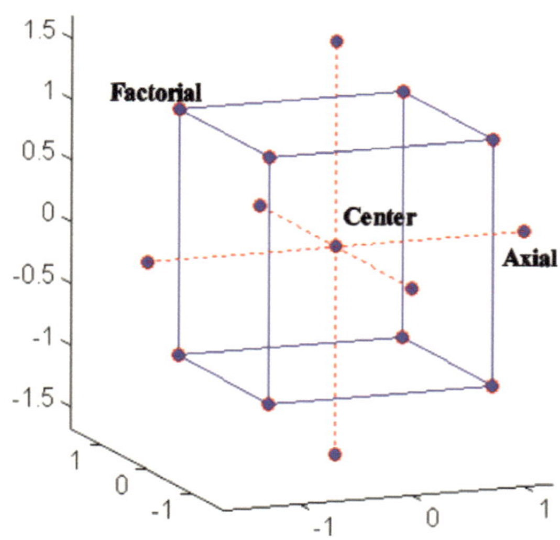

Figure 3. Design of experiment using central composite design.

Level ->	Lowest	Low	Center	High	Highest
Coded value (x)	−1.682	−1	0	1	1.682
Cutting speed V_c (m/min) turning	69.9	90.4	120	150	171.4
Feed rate f (mm/min) turning	55.6	72	96	120.6	136.6
Depth of cut a_p (mm) milling	1.83	2.0	2.5	2	2.67

Table 1. Level of cutting parameters used for central composite design.

was performed using MINITAB 17 statistical software. For the present work, based on number of input factor k, the value of α was taken as 1.682. The coded and natural levels of the independent variables for design of experiments are presented in **Table 1**. Five levels of cutting parameters were calculated in central composite design using Eq. (12) shown above. After defining levels of cutting parameters, sequence of experiments were generated using MINITAB 17 statistical software using central composite design for turning and milling operations. **Table 2** shows the 20 sets of experiment in terms of coded values of cutting parameters sequenced according to run order. The number of experiments was generated based on number of input factors and their levels.

$$x_1 = \frac{V_c - (V_{cmax} + V_{cmin})/2}{(V_{cmax} - V_{cmin})/2}; x_2 = \frac{f - (f_{max} + f_{min})/2}{(f_{max} - f_{min})/2}; x_3 = \frac{a_p - (a_{pmax} + a_{pmin})/2}{(a_{pmax} - a_{pmin})/2} \tag{12}$$

Where x is coded value of level of individual cutting parameter, V_c is cutting speed in m/min, f is feed rate in mm/rev, a_p = depth of cut in mm.

Std order	Run order	Pt type	Blocks	Cutting speed	Feed rate	Depth of cut
5	1	1	1	−1	−1	1
6	2	1	1	1	−1	1
4	3	1	1	1	1	−1
14	4	−1	1	0	0	1.681793
1	5	1	1	−1	−1	−1
2	6	1	1	1	−1	−1
19	7	0	1	0	0	0
20	8	0	1	0	0	0
7	9	1	1	−1	1	1
3	10	1	1	−1	1	−1
9	11	−1	1	−1.68179	0	0
10	12	−1	1	1.681793	0	0
8	13	1	1	1	1	1
17	14	0	1	0	0	0
11	15	−1	1	0	−1.68179	0
18	16	0	1	0	0	0
15	17	0	1	0	0	0
16	18	0	1	0	0	0
13	19	−1	1	0	0	−1.68179
12	20	−1	1	0	1.681793	0

Table 2. Sequence of experiments obtained using MINITAB.

In the present case study, minimization of surface roughness is done for turning and milling operations. Surface roughness was measured for each machining operation. In order to compensate measuring error, surface roughness was measured at three locations on the machined surface and average value is taken. **Table 3** show the list of experiments and corresponding surface roughness in turning operations.

Second order models are developed for surface roughness in turning using RSM. After developing models, ANOVA analysis is done to see significant and insignificant terms in the models as shown in **Table 4**. Insignificant terms are identified and eliminated using backward elimination procedure. In **Table 4**, the variable for which the value of 'p' is less than 0.05 indicates that the term in the model has a significant effect on the response.

The ANOVA results shown in **Table 4** demonstrate that the model is highly significant, and the lack of fit is non-significant. Model showed a correlation coefficient (R^2) of 93.13% for turning which means more than 90% of the data can be explained by these models. Furthermore, the significance of each coefficient in the full model was examined by the F-values and p-values. Larger values of "F" and smaller values of p ($p < 0.1$) indicate that the corresponding variable

Run type	Cutting speed V_c (m/min)	Feed rate f (mm/min)	Depth of cut a_p (mm)	Surface roughness R_a (μm)
Center	120.6	96	1.5	0.541
Center	120.6	96	1.5	0.559
Axial	69.9	96	1.5	0.819
Factorial	150.8	72	1	0.457
Axial	120.6	96	2.34	0.608
Factorial	90.4	120	1	0.766
Factorial	150.8	120	1	0.483
Center	120.6	96	1.5	0.592
Center	120.6	96	1.5	0.592
Factorial	150.8	72	1	0.404
Factorial	150.8	120	2	0.474
Axial	120.6	136.4	1.5	0.602
Center	120.6	96	1.5	0.583
Axial	120.6	96	0.66	0.533
Factorial	90.4	72	2	0.747
Factorial	90.4	120	2	0.844
Center	120.6	96	1.5	0.582
Axial	120.6	55.6	1.5	0.554
Axial	171.4	96	1.5	0.386
Factorial	90.4	72	1	0.747

Table 3. Surface roughness measurement after turning operation.

Source	Sum of square	DF	Mean of square	F value	p value Prob > F
Model	0.31	9	0.035	29.64	<0.0001
V_c	0.30	1	0.30	253.51	<0.0001
f	6.283e-3	1	6.283e-3	5.36	0.0432
a_p	4.499e-3	1	4.499e-3	3.84	0.0786
$V_c \times f$	4.572e-5	1	4.572e-5	0.039	0.8474
$V_c \times a_p$	1.591e-4	1	1.591e-4	0.14	0.7203
$a_p \times f$	2.841e-5	1	2.841e-5	0.024	0.8794
V_c^2	3.859e-3	1	3.859e-3	3.29	0.0998
f^2	8.258e-4	1	8.258e-4	0.70	0.4211
a_p^2	3.612e-4	1	3.612e-4	0.31	0.5912
Residual	0.012	10	1.173e-3		
Lack of fit	9.587e-3	5	1.917e-3	4.47	0.629
Pure error	2.143e-3	5	4.286e-4		
Core total	0.32	19			

Table 4. ANOVA analysis for surface roughness as response and cutting parameters as variables in turning operation.

is highly significant. Hence, the results given in **Table 4**, suggest that the influence of f^2 (square of feed rate), a_p^2 (square of depth of cut), $V_c \times f$ (cutting speed × feed rate), $V_c \times a_p$ (cutting speed × depth of cut), and $f \times a_p$ (feed rate × depth of cut) are non-significant and therefore, can be removed from the full model to further improve the mode as shown in Eq. (13).

$$R_a = 1.27686 - 0.000897964 \times V_c + 0.0008937 \times f + 0.036303 \times a_p + 1.69203e - 7 \times V_c^2 \quad (13)$$

4.1. Validation of developed model for surface roughness in turning operation

In order to verify the adequacy of the model developed, five validation experiments were performed as depicted in **Table 5**. The conditions were those which have not been used previously but are within the range of the levels defined previously. The predicted values from

Parameters	Exp. 1	Exp. 2	Exp. 3	Exp. 4	Exp. 5
Cutting speed (m/min)	72.5	95.0	110.0	130.0	160.0
Feed rate (mm/min)	60	80	100	125	140
Depth of cut (mm)	0.8	0.9	1.4	1.6	1.8
Predicted R_a (μm)	0.797	0.681	0.634	0.565	0.464
Actual R_a (μm)	0.772	0.748	0.721	0.6325	0.501
% Error	−3.18	9.96	13.69	11.95	8

Table 5. Confirmation experiments for validating surface roughness model for turning operation.

the equation developed for surface roughness and the actual experimental value were compared. The percentage errors were calculated. All these values are presented in **Table 5**. The percentage error range between the actual and predicted value is −3.18 to 13.69% which is acceptable. Residual from the least square fit is defined by $e_i = y_i - y^*$ for i = 1, 2,….20 where y_i is the observed response (Surface roughness) and y^* is the predicted response. A check of the normality assumption may be made by constructing a normal probability plot of the residuals. If the residuals plot is approximately along a straight line, then the normality assumption is satisfied. **Figure 4** presents a plot of residuals e_i versus the predicted response y^* and it reveals no apparent problem with normality.

From the confirmation experiments and normal probability plot of residual, it is observed that the developed model can predict the surface roughness in turning operation. **Figure 5** shows the

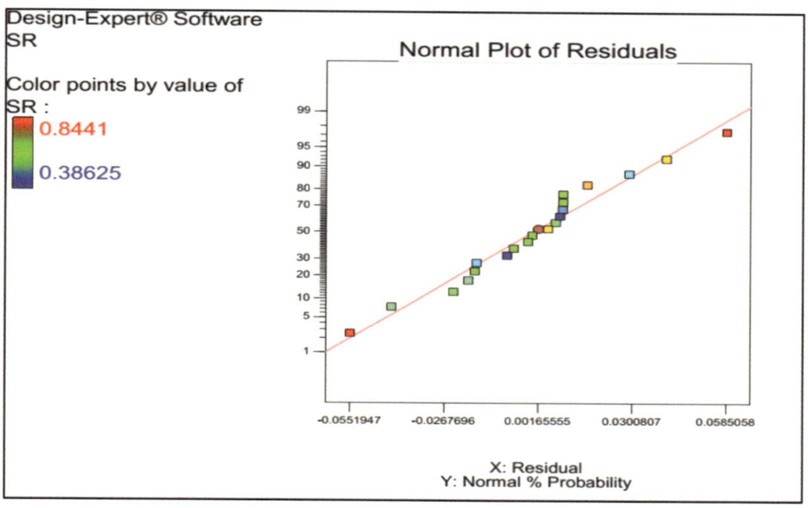

Figure 4. Normal probability plot of residual for surface roughness in turning operation.

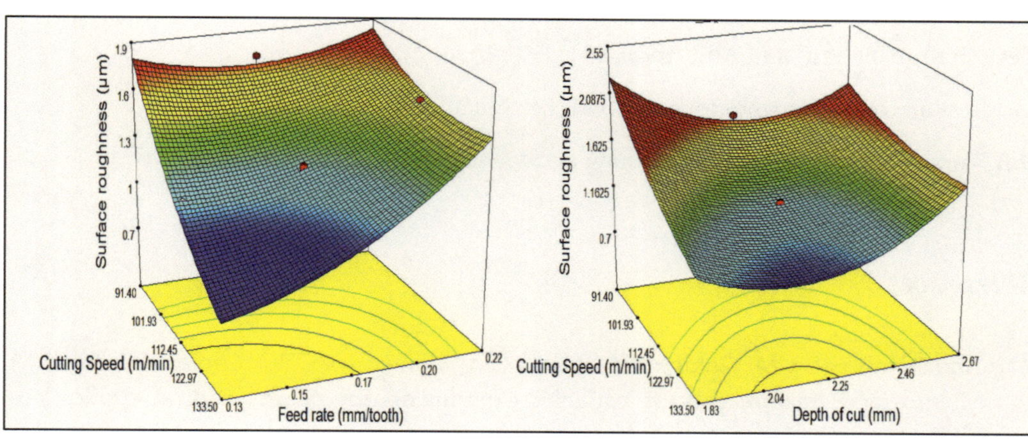

Figure 5. Response surface plots.

response surface plots give a graphical display of these quantities. Typically, the variance of the prediction is also of interest, because this is a direct measure of the likely error associated with the point estimate produced by the model.

From the response surface plots also, it is observed that interaction of cutting speed/feed rate is strongly affecting the surface roughness value whereas interaction of feed/ doc and cutting speed/doc has negligible effect over surface roughness [22, 23].

5. Summary

From the above study it can be concluded that experimenter can predict the response using proper design of experiment where proper underlying mechanism of the process is not fully understood. Proper fitting of response from experimental data can be done by design of experiment, regression modeling technique, statistical analysis and optimization. Following conclusions can be made based on the case study:

- Design of experiments is a very structured methodology for planning and designing a sequence of experiments.
- Analysis of variance (ANOVA) was used to identify significant input variables for particular response.
- Prediction model can be developed for a response with correlation coefficient more than 90% which confirm that the models properly explain the experimental data.
- The developed predictive model can help industries in achieving appropriate output for improving productivity.

Author details

Neelesh Kumar Sahu* and Atul Andhare

*Address all correspondence to: neeleshmecher@gmail.com

Shri Ramdeobaba College of Engineering and Management, Nagpur, India

References

[1] Anderson-Cook CM, Goldfarb H, Borror CM, Montgomery DC, Canter KG, Twist JA. Mixture and mixture process variables experiments for pharmaceutical applications. Pharmaceutical Statistics. 2004;**3**:247-260

[2] Chung PJ, Goldfarb HB, Montgomery DC. Optimal designs for mixture-process experiments with control and noise variables. Journal of Quality Technology. 2007;**39**:179-190

[3] Mosteller F, Tukey JW. Data Analysis and Regression: A Second Course in Statistics. Addison-Wesley Series in Behavioural Science: Quantitative Methods. Boston, MA, USA: Addison-Wesley Publishing Company; 1977

[4] Myers RH, Montgomery DC, Anderson-Cook CM. Response Surface Methodology: Process and Product Optimization Using Designed Experiments. New Jersey, USA: John Wiley & Sons; 2016

[5] Choudhary AK, Chelladurai H, Kannan C. Optimization of combustion performance of bioethanol (water hyacinth) diesel blends on diesel engine using response surface methodology. Arabian Journal for Science and Engineering. 2015;**40**(12):3675-3695. DOI: 10.1007/s13369-015-1810-y

[6] El-Tayeb NSM, Yap TC, Venkatesh VC, Brevern PV. Modeling of cryogenic frictional behaviour of titanium alloys using Response Surface Methodology approach. Materials & Design. 2009;**30**(10):4023-4034. DOI: 10.1016/j.matdes.2009.05.020

[7] Condra LW. Reliability Improvement with Design of Experiments. New York: Marcel Dekker; 1993

[8] Cornell JA. Experiments with Mixtures, Designs, Models, and the Analysis of Mixture Data. 3rd ed. New York: John Wiley & Sons; 2002

[9] Allen TT, Yu L, Schmitz J. An experimental design criterion for minimizing meta-model prediction errors applied to a die casting process design. Applied Statistics. 2003;**52**:103-117

[10] Montgomery DC. Design and Analysis of Experiments. New Jersey, USA: John Wiley & Sons; 2017

[11] Drain D, Carlyle WM, Montgomery DC, Borror CM, Anderson-Cook CM. A genetic algorithm hybrid for constructing optimal response surface designs. Quality and Reliability Engineering International. 2004;**20**:637-650

[12] Draper NR. Center points in second-order response surface designs. Technometrics. 1982;**24**:127-133

[13] Andrews DF. A robust method for multiple linear regression. Technometrics. 1974;**16**:523-531

[14] Bartlett MS, Kendall DG. The statistical analysis of variance heterogeneity and the logarithmic transformation. Journal of the Royal Statistical Society, Series B. 1946;**8**:128-150

[15] Cornell JA. Fitting models to data from mixture experiments containing other factors. Journal of Quality Technology. 1995;**27**(1):13-33

[16] Ulutan D, Ozel T. Machining induced surface integrity in titanium and nickel alloys: A review. International Journal of Machine Tools and Manufacture. 2011;**51**(3):250-280

[17] Che-Haron CH, Jawaid A. The effect of machining on surface integrity of titanium alloy Ti–6% Al–4% V. Journal of Materials Processing Technology. 2005;**166**(2):188-192. DOI: 10.1016/j.jmatprotec.2004.08.012

[18] Razfar MR, Asadnia M, Haghshenas M, Farahnakian M. Optimum surface roughness prediction in face milling X20Cr13 using particle swarm optimization algorithm. Proceedings

of the Institution of Mechanical Engineers, Part B: Journal of Engineering Manufacture. 2010; **224**:1645-1653

[19] Sharif S, Mohruni AS, Noordin MY, Venkatesh VC. Optimization of surface roughness prediction model in end milling Titanium Alloy (Ti-6Al4V). In: Proceeding of ICOMAST2006, International Conference on Manufacturing Science and Technology, 2006. Melaka, Malaysia: Faculty of Engineering and Technology, Multimedia University; 2006. pp. 55-58

[20] Mukherjee I, Ray PK. A review of optimization techniques in metal cutting processes. Computers & Industrial Engineering. 2006;**50**(1–2):15-34. DOI: 10.1016/j.cie.2005.10.001

[21] Box GEP. The effect of errors in the factor levels and experimental design. Technometrics. 1963;**6**:247-262

[22] Sahu NK, Andhare AB. Optimization of surface roughness in turning of Ti-6Al-4V Using Response Surface Methodology and TLBO. In: ASME 2015 International Design Engineering Technical Conferences and Computers and Information in Engineering Conference. New York USA: American Society of Mechanical Engineering; 2015

[23] Sahu NK, Andhare AB. Modelling and multiobjective optimization for productivity improvement in high speed milling of Ti–6Al–4V using RSM and GA. Journal of the Brazilian Society of Mechanical Sciences and Engineering. 2017;**39**(12):5069-5085

Chapter 3

Design of Experiments Applied to Antibiotics Degradation by Fenton's Reagent

André Luís de Castro Peixoto,
Ademir Geraldo Cavallari Costalonga,
Mateus Nordi Esperança and
Rodrigo Fernando dos Santos Salazar

Additional information is available at the end of the chapter

http://dx.doi.org/10.5772/68097

Abstract

Advanced oxidation technologies (AOTs) are processes affected by a large number of parameters such as iron (Fe^{2+}) and H_2O_2 concentrations, pH, temperature, light intensity and chemical composition (organics and inorganics). In addition, for different industrial chemical processes, there are different effluents, which greatly vary in chemical composition from each other, not allowing a unique approach for all of them. Thus, it is necessary to adjust AOT parameters to the specific effluent to be treated. In such context, statistical design of experiments (DoE) and response surface methodology (RSM) emerge as important and widely used tools to determine the effects of multiple variables on wastewater treatment processes such as photo-Fenton. A revision of academic studies based on degradation of antibiotics-containing effluents is presented. The chapter also presents commercial cases of AOT and electrical efficiency considerations.

Keywords: design of experiments, planning of experiments, optimization, advanced oxidation technology, Fenton's reagent, pharmaceutic compounds, full scale, figures-of-merit electrical efficiency

1. Introduction

Analgesics, hormones, anaesthetics, mainly antibiotics are pharmaceutical compounds that can lead to damages to the environment and public health when are improperly discarded. The main sources of contamination by compound drugs are due to wastewater, effluent and inadequate disposal by pharmaceutical industries. Many studies related the presence of these compounds in surface water, groundwater, tap water and urban sewage samples. On the other hand, social and legal demands on the impact of chemical effluents on the environment have driven the development of many procedure and technologies in order to treat wastewater and poisoning water by the presence of these compounds.

By this way, those technologies based on advanced oxidation technologies (AOTs) have been extensively studied for the decomposition of a series of persistent and recalcitrant compounds present in wastewater into less toxic and/or biodegradable substances. In many cases, these technologies result in the formation of CO_2, H_2O and inorganic compounds of all non-oxygen heteroatom from the oxidation of the organic matter present in the persistent compounds. The AOTs were defined by Glaze et al. [1] as physicochemical processes involving the generation of transient species of high oxidizing power, among which the hydroxyl radical (OH) stands out [1]. This radical has a high oxidizing power ($EPH_{HO\bullet/HO}$ +2.8V, 25°C) and can be generated by photons (including sunlight) or by other forms of energy, being able to mineralize organic pollutants to non-toxic as CO_2 and H_2O. Some AOTs, such as heterogeneous photocatalysis, radiolysis and other advanced techniques, allow the transformation of toxic contaminants that are not susceptible to oxidation, such as metal ions and halogenated compounds [2]. Among the most studied AOTs, heterogeneous photocatalysis employing semiconductors and/or H_2O_2/UV/semiconductor has played an important role in relation to emerging technologies for water treatment, due to the large number of investigations on the subject compared to other AOTs studied, which is most extensively presented in the literature [2, 3].

The catalytic oxidation of tartaric acid in the presence of ferrous salts and hydrogen peroxide was reported by Fenton in the mid-1890s. The oxidation of organic compounds under UV irradiation in the presence of ferric ion in acidic media was verified in the 1950s when it was postulated that the electron transfer initiated by the irradiation resulted in the generation of $^{\bullet}OH$, responsible for the oxidation reactions [4]. Degradation efficiency of different classes of toxic organic compounds has made the photo-Fenton process activated by sunlight quite attractive, investigating the use of AOT based on Fenton processes, Photo-Fenton, as well as derived processes for different applications of environmental interest. In this way, the degree of interest has led to the rapid technological development of this class of AOT. Consequently, it is possible to find these pilot scale systems being tested for the final treatment of water for supply and as tertiary stage of municipal sewage treatment in Canada and Spain [5, 6].

As scientific and technological advances are made involving the use of different advanced oxidative technologies for effluent and wastewater remediation, the need to optimize these processes in order to be commercially available for the mineralization and stabilization of recalcitrant compounds is growing. In this sense, statistical tools based on chemometrics and design of experiments (DoE) have been used to evaluate the figures of merit to extend the range of commercially available systems. Chemometrics is defined as the application of mathematical and statistical models and methods for the solution of chemical problems, in order to maximize the data collection and to allow the extraction of useful information from the obtained data. The development of analytical equipment and chemical processes has led to a need for advances in experimental design methods, with the objective of obtaining secure information in a shorter time span for instrumental calibration and process efficiency analysis [7, 8].

Design of experiments (DoE) is defined as a set of statistical techniques applied to the planning, conduction, analysis and interpretation of controlled tests to find and define factors that influence the values of a parameter or a group of parameters. Its basic principle allows varying all the levels of all the variables, discrete or continuous, in a programmed and rational way, reducing the number of experiments without limiting the number of factors to be analysed. The use of complete factorial design becomes necessary when assessing the influence of variables, without running the risk of excluding factors or interactions that may be important [8]. Although for economical reasons, fractional factorial design is usually applied [7, 8].

However, only a small fraction of the scientific papers related to advanced oxidative technologies makes use of the electrical energy consumption figure-of-merit (EE/O). In our understanding, one of the best response variables that could be used in the study of wastewater degradation (pharmaceuticals or not) would be EE/O. With EE/O as variable response in a design of experiments, not only the main factors associated to the advanced oxidative processes (light source, catalyst, H_2O_2, O_3, Fe^{2+} etc.) but also the kinetics of the reaction and the energy cost would be used in an experimental study to reach the objective of the process. There is a gap in the scientific community regarding the use of the EE/O tool that could be explored by researchers in conjunction with the traditional statistical tool design of experiments.

In this sense, the aim of this chapter is to present the use of physicochemical systems based on advanced oxidation technology (AOT) for the effluents remediation containing pharmacological residues. In addition, it is intended to present the state-of-the-art about design of experiments (DoE) and other statistical tools in order to evaluate the figures-of-merit of photochemical reactors for the treatment of pharmaceutical effluent.

2. Basics of DoE

Design of experiments (DoE) is not a set of formulas but a technique used to plan experiments, behind there is a theoretical support with good reasoning math. Basically, it looks good statistical accuracy in response and a lower cost. Therefore, it is a technique of great importance for the industry and its use permits the obtaining of more reliable results and thus saving time and resources.

Design of experiments, in statistics, refers to the whole area of studies that develops technical planning and analysis of experiments. The main techniques of planning experiments are old; however, the great many of these techniques requires a good amount of calculations being essential to use the computational resources.

An experiment is nothing else than a procedure or test in which deliberate alterations are made in input variables of a process to observe, identify and evaluate possible alterations in the response variable, as well as the reasons of such alteration. DoE is a test, or series of tests in which the changes in input variables of a process are known, for then, observe which changes occur in the response variable.

Why plan an experiment? To have a set of rules or way outline to obtain a mathematical model that adequately describes the process investigated, using as few experiments as possible. The planning will let efficiency and economy in the process.

The planning of an experiment investigates potential factors whose variation might impact the response-variable (process output). The planning will be used to obtain valid results and introduce objective conclusions. The planning must maximize the quantity of information that can be obtained for each variation performed. The experimental factor is the variable that is controlled to check its effect in the response; the factors can be classified into two types: qualitative or quantitative. A proper planning has to take into consideration some items: a DOE investigates a list of potential factors whose variation might impact the process output

- Recognition of the problem
- Set experiment objectives
- Define and know the resources
- Optimize resources to meet the objective
- Carry out the experiment and analysis of results

2.1. Recognition of the problem

The Fenton's process consists in the use of H_2O_2/Fe^{2+} and was first observed by Fenton in 1894. In solution, ferrous ions (Fe^{2+}) initiate and catalyze the decomposition of H_2O_2 and lead to the formation of hydroxyl radical ($^{\bullet}OH$). Mixtures of Fe^{2+} and H_2O_2 are called Fenton reagent. If Fe^{2+} is replaced by Fe^{3+}, it is called Fenton-like reagent.

The efficiency of this process is directly linked or related to other experimental parameters, such as pH, concentration of hydrogen peroxide, iron concentration, concentration of organic matter, intensity of radiation UV-Vis, exposure time and volume of solution.

2.2. Set the experiment objectives

A good problem implies to define the goal of the experiment. This objective must be non-biased, specific, measurable and should lead to a practical result. The aim is to answer the following question: What do you want to investigate? It depends on the operational variables (factors) to be studied, and the way they relate to each other (synergism or antagonism effects) in the process. The DoE allows evaluating synergic and antagonistic effects of the operational variables simultaneously and with a reduced number of experiments. Therefore, one must investigate the changes in the input variables of this process and then observe which changes occur in the response variable. The objective of experimental design is to optimize the experimentation process to obtain as much information as possible.

2.3. Define and know the resources

One must choose and select the variables that are possible to be studied and which probably interfere with the system. In the photo-Fenton process, the variables most studied are pH, concentration of hydrogen peroxide, the concentration of iron, concentration of organic matter, intensity of radiation (UV-Vis) and time of exposure. The variable response must be chosen with the assurance that the result really supplies useful information about the process under study. One of the possible variables was chosen is the Total Organic Carbon (TOC) removal rate.

The choice of factors and levels must encompass the ranges over which these factors will vary and the specific levels in which each assessment will be carried out. The variables that were not selected must be fixed during the entire experiment. Instruments, equipment, places, people, reagents, time spent and cost of the process are external variables to the process that may influence the application of the method after optimization.

2.4. Optimize resources to meet the objective

Optimization of resources (e.g. chemicals, materials, energy and staff) is a mandatory part of a process. In this step, the sample size should be defined and which and how many levels will be used for each factor should also be known. Only factors that affect the response variable should be selected. Moreover, the way the measures be carried out, the data acquisition method, the equipment or instruments that will be necessary along the experiments should also be defined. Assess whether the parameters involved can be harmed, and evaluate what can go wrong (time, costs, company reputation, etc.). In the optimization of resources step, even in a preliminary way, choose which statistical method will be used to evaluate results. The choice of DoE involves consideration of sample size (number of replications), selection of an appropriate order of experimental runs, formation of experimental blocks or other restrictions involved.

Plan experiments is to define an experimental data acquisition sequence to reach certain objective, among the methods the factorial planning is most useful when the user wants to study the effects of two or more variables, so that in each attempt, or replies, all possible combinations of the levels of each variable are investigated.

2.5. Carry out the experiment and analysis of results

It is necessary to carry out the experiments with rigor—paying attention to minimum details in each run—to ensure that everything is being performed according to the planning. Any errors in the experimental procedure disable the experiment. In data analysis, statistical methods must be used so that the results and conclusions are not an opinion but goals. If the experiment was planned properly and was carried out in agreement to the planning, the application of statistical methods would not be complicated. Excellent statistical packages exist to help in data analysis and graphical methods are the simplest and easiest in data interpretation.

2.5.1. Software

Statistica—developed by Stat Soft—it is one of the systems with an easy-to-use look with a good graphical interface, the software can make correlations, brings several descriptive statistics, a wide range of tables and a variety of graphical analyses. http://www.statsoft.com/

SAS—'Statistical Analysis System' Developed by SAS—It is one of the best systems used by statisticians. SAS is an integrated system for data analysis applications, consisting of data recovery, file management, statistical analysis, database access, generation of graphs and generation of reports (http://www.sas.com/).

MINITAB—developed by Minitab—this is a classic software for statistical purposes. Its interface is similar to a spreadsheet such as Microsoft Excel or Calc of OpenOffice/LibreOffice but with the ability to perform complex statistical analyses. It offers tools of Quality Control, Planning of Experiments (DoE), Reliability Analysis and General Statistics. http://www.minitab.com/

ACTION STAT—developed by Estatcamp. Action Stat under R—which is a free software environment for statistical computing and graphics, and one of the most widely used statistical environments. Action Stat connects with Excel to provide a graphical interface for statistical applications (http://www.portalaction.com.br/).

DESIGN EXPERT—Developed by Stat-Ease—it is a Windows®-based software intended to optimize a product or process. It provides many statistical tools such as: two-level factorial screening designs; general factorial studies; response surface methods (RSM); mixture design techniques; combinations of process factors, mixture components and categorical factors and design and analysis of split plots (http://www.statease.com).

3. Design of experiments applied to advanced oxidation technologies

Advanced oxidation technologies are affected by a large number of parameters such as iron (Fe^{2+}) and H_2O_2 concentrations, pH, temperature, light intensity, organic chemical content, among others. In addition, for different industrial chemical processes there are different effluents, which greatly vary in this chemical composition from each other, not allowing a unique approach for all of them. Thus, it is necessary to adjust AOT parameters to the specific effluent to be treated.

This adjustment could be performed using the one-variable-at-a-time (OVAT) approach, but this procedure is time-consuming and less effective due to multiple variable nature of AOT. Statistical design of experiments (DoE) and response surface methodology (RSM) emerge as important and widely used tools to determine the effects of multiple variables on objective functions to be optimized. Different types of DoE used for AOT evaluation include two-level factorial design (2^k FD), central composite design (CCD) and Box-Behnken design (BBD), as can be seen on **Table 1**.

3.1. Two-level factorial design (2^k FD)

Factorial designs are a widely used and efficient way to evaluate the effects of two or more factors and the interactions among them on response variables. When compared to the one-variable-at-a-time (OVAT) experiments, factorial designs exhibit higher relative efficiency, avoid misleading conclusions when interactions are present and allow the effect estimation of

Reference	AOT	Organic matter	DoE	Independent factor: range
Ay and Kargi [9]	Fenton	Amoxicillin	BBD	Amoxicillin concentration: 10–200 mg L^{-1} $[H_2O_2]$: 10–500 mg L^{-1} $[Fe^{2+}]$: 0–50 mg L^{-1}
Ay and Kargi [10]	Photo-Fenton	Amoxicillin	BBD	Amoxicillin concentration: 10–200 mg L^{-1} $[H_2O_2]$: 10–500 mg L^{-1} $[Fe^{2+}]$: 0–50 mg L^{-1}
Irani et al. [11]	Coupled adsorption/photo-Fenton	Phenol and paracetamol	BBD	pH: 3–4 Phenol initial concentration: 20–100 mg L^{-1} Paracetamol initial concentration: 20–100 mg L^{-1} NaX to cobalt ferrite nanoparticles ratio: 0.5–1.5
Diniz [12]	Fenton	Hospital sewage	2^2 FD	$[Fe^{2+}]$: 0.1–0.5 g L^{-1} $[H_2O_2]$: 0.1–0.5 g L^{-1}
Marcelino [13]	Ozonation	Amoxicillin and cephalexin	2^2 FD	pH: 5–12 O_2 flow rate: 0.5–1 $Lmin^{-1}$
Marcelino [13]	Fenton	Amoxicillin and cephalexin	2^2 FD	$[Fe^{2+}]$: 100–500 mg L^{-1} $[H_2O_2]$: 1000–1500 mg L^{-1}
Marcelino [13]	Photo-Fenton	Amoxicillin and cephalexin	2^2 FD	$[Fe^{2+}]$: 100–500 mg L^{-1} $[H_2O_2]$: 1000–1500 mg L^{-1}
Dwivedi et al. [14]	Fenton	Carbamazepine	2^3 FD	pH: 2–6 $[H_2O_2]$: 0.5–3 g L^{-1} Contact time: 10–30 min
Silva [15]	Photo-Fenton	Amoxicillin	2^4 FD	Amoxicillin concentration: 20–60 mg L^{-1} $[Fe^{2+}]$: 5–15 mg L^{-1} $[H_2O_2]$: 50–150 mg L^{-1} UV light intensity: 0–96 W
Dwivedi et al. [14]	Fenton	Carbamazepine	CCD	pH: 1.37–5.62 $[H_2O_2]$: 0.34–11.65 g L^{-1}
Pérez-Moya et al. [16]	Photo-Fenton	Sulfamethazine	CCD	$[H_2O_2]$: 176–1,024 mg L^{-1} $[Fe^{2+}]$: 12–68 mg L^{-1}
Silva et al. [17]	Photo-Fenton	Phenol	CCD	[NaCl]: 0.04–5,857.86 ppm

Reference	AOT	Organic matter	DoE	Independent factor: range
				$[Na_2SO_4]$: 0.04–5,857.86 ppm
Affam et al. [18]	Modified Fenton ($FeGAC/H_2O_2$)	Amoxicilin and cloxacillin	CCD	H_2O_2/COD molar ratio: 0.32–3.68
				[FeGAC]: 1–6 g L^{-1}
				Reaction time: 39.55–140.45 min
Almeida [19]	Photoelectro-Fenton	Paracetamol	CCD	Electric current: 3.5–8.5 A
				$[Fe^{2+}]$: 3.30×10^{-4}–1.17×10^{-3} mol L^{-1}
				pH: 1.37–4.36
Frade [20]	Fenton	Enrofloxacin	CCD	$[H_2O_2]$: 0.1–0.9 g L^{-1}
				$[Fe^{2+}]$: 5.0×10^{-3}–1.2×10^{-1} g L^{-1}
				Temperature: 0–40°C
Homem et al. [21]	Fenton		CCD	$[H_2O_2]$: 4.2×10^{-4}–4.28×10^{-3} g L^{-1}
		Amoxicillin		$[Fe^{2+}]$: 3.0×10^{-5}–3.5×10^{-4} g L^{-1}
				Temperature: 20–70°C
Rozas et al. [22]	Fenton and photo-Fenton	Ampicillin	CCD	$[Fe^{2+}]$: 5.3×10^{-5}–8.7×10^{-5} mol L^{-1}
				$[H_2O_2]$: 2.3×10^{-4}–5.7×10^{-4} mol L^{-1}
				pH: 2.3–5.7
Sarrai et al. [23]	Photo-Fenton	Tylosin	CCD	$[Fe^{2+}]$: 6.4×10^{-4}–7.36×10^{-3} g L^{-1}
				$[H_2O_2]$: 1.32×10^{-4}–4.68×10^{-4} g L^{-1}
				pH: 1.89–3.9
Zaidan et al. [24]	Photo-Fenton	Phenol	CCD	$[Fe^{2+}]$: 1.59×10^{-3}–1.50×10^{-2} mol L^{-1}
				$[H_2O_2]$: 5.93–21.06 g L^{-1}
				Reaction time: 39.55–140.55 min
Arslan-Alaton et al. [25]	Photo-Fenton-like	Naphthalene sulphonic acid (H-acid)	CCD	Reaction time: 6–30 min
				[COD]: 0.15–0.75 g L^{-1}
				$[H_2O_2]$: 1×10^{-2}–5×10^{-2} mol L^{-1}
				$[Fe^{3+}]$: 5.0×10^{-4}–2.5×10^{-3} mol L^{-1}

Reference	AOT	Organic matter	DoE	Independent factor: range
Domínguez et al. [26]	Integrated Fenton/photo-Fenton-like	Carbamazepine	CCD	pH: 2.5–4.5 $[H_2O_2]$: 0–1.68 × 10^{-4} mol L^{-1} $[Fe^{2+}]$: 0–1.68 × 10^{-5} mol L^{-1} $[Fe^{3+}]$: 0–1.68 × 10^{-5} mol L^{-1}

Table 1. Studies applying design of experiments for evaluation of advanced oxidation technologies.

a factor at different levels of the other factors. The two-level factorial design with k factors (2^k) is the most important of these special cases. In this methodology, the factors vary from a 'low' level (also designated as −1) to a 'high' level (also designated as +1). The 2^k factorial design (2^k FD) is a powerful factor screening tool at initial research stages, since it provides the smallest number of experiments to be run when many factors are investigated [27].

The number of runs in a set of two level factorial design comprises 2^k factorial points and central points. A usual matrix (treatment combinations) for two (2^2) and three (2^3) factors are presented in **Table 2**, in a coded representation.

In **Table 2**, runs 1–4 (for 2 factors, **Table 2a**) and 1–8 (for 3 factors, **Table 2b**) are called factorial points and consists in all 2^k possible combinations for the low and high levels of all factors. Runs 5–7 (for 2 factors, **Table 2a**) and 9–11 (for 3 factors, **Table 2b**) are called central points and are used to obtain an estimate of the error, in order to allow the identification of the significant factors for a defined confidence interval. In general, three to five central points are recommended to obtain a good estimation of response variance. At the central points, the factors assume the mean value between their own low and high levels.

Geometrically, the space delimitated by the factors range of variation is represented by a square and a cube, as can be seen in **Figure 1**, for two and three factors, respectively.

The results of 2^k FD can be expressed using a first-order regression model (Eq. (1)):

$$y = \beta_0 + \sum_{j=1}^{k} \beta_j x_j + \sum\sum_{i<j} \beta_{ij} x_i x_j \tag{1}$$

Where y corresponds to the response variable, x_j (x_i) represents the coded factors, β_0 is the mean value of response variable, β_j's represent the linear coefficients and β_{ij}'s represent the interaction coefficients. The relationship between the coded (x) and natural (actual) factors (X) is as follows:

$$x = \frac{X - X_0}{\Delta} \tag{2}$$

$$\Delta = \frac{X_{\text{high}} - X_{\text{low}}}{2} \tag{3}$$

Run	Factor A	Factor B
1	−1	−1
2	+1	−1
3	−1	+1
4	+1	+1
5	0	0
6	0	0
7	0	0

(a)

Run	Factor A	Factor B	Factor C
1	−1	−1	−1
2	+1	−1	−1
3	−1	+1	−1
4	+1	+1	−1
5	−1	−1	+1
6	+1	−1	+1
7	−1	+1	+1
8	+1	+1	+1
9	0	0	0
10	0	0	0
11	0	0	0

(b)

Table 2. Two-level factorial designs: (a) two factors; (b) three factors.

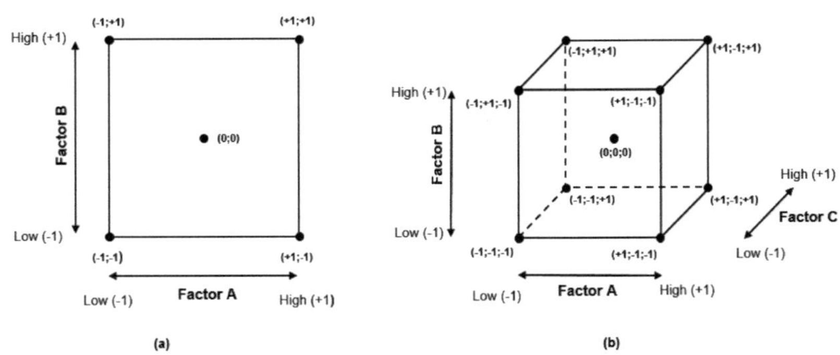

Figure 1. Geometric view of two-level factorial designs: (a) two factors and (b) three factors.

Where X_0, X_{low}, and X_{high} are the value of the natural factor at central point, low level and high level, respectively. For the cases discussed above, Eq. (1) becomes:

$$k = 2\textit{factors}: y = \beta_0 + \beta_1 x_1 + \beta_2 x_2 + \beta_{12} x_1 x_2 \quad (4)$$

$$k = 3\textit{factors}: y = \beta_0 + \beta_1 x_1 + \beta_2 x_2 + \beta_3 x_3 + \beta_{12} x_1 x_2 + \beta_{13} x_1 x_3 + \beta_{23} x_2 x_3 \quad (5)$$

In Eq. (1), coded variables are preferably used instead of natural factors since coded factors allow an effective evaluation of relative size of factors effects. This means that depending on the ranges and units of natural variables, their relative effects could be masked, leading to erroneous simplification of the model.

The general approach to the statistical analysis of 2^k factorial design consists in the following steps [27]:

- Obtain the generalized model (full model) adjusting the regression model described by Eq. (1) to the experimental data.
- Define a confidence interval and perform an Analysis of Variance (ANOVA) to identify the statistical significant terms of Eq. (1) (single factors and interaction factors).
- Refine the model, excluding the non-significant terms from the model and adding them to the lack of fit, and recalculate the coefficients.
- Verify the model adequacy performing a residual analysis.
- Built response surfaces (or contour plots) to perform the graphical interpretation of the results.

Since Eq. (1) is a first-order model, the response surface described is a plane. Then, it is not possible to affirm that the highest value exhibited corresponds to an optimal value. However, the response surface can be used to obtain a direction of potential improvement using the method of steepest ascent.

3.2. Central composite design (CCD)

The model described by Eq. (1) allows the representation of some curvature on response surface, as the result of twisting of the plane caused by the interaction of factors. However, if there is no curvature with identification of maximum or minimum points in the graph, it is then necessary to provide more information for the mathematical model. For that cases, a second-order model regression may be used [27].

If there is no curvature, then the mean response at the centre point equals the average of the mean response of the factors at their low and high settings (the corners of the design space). Curvature is detected when the average mean response at the centre points is significantly greater or less than the average mean response of the factors at their low and high settings.

$$y = \beta_0 + \sum_{j=1}^{k} \beta_j x_j + \sum\sum_{i<j} \beta_{ij} x_i x_j + \sum_{j=1}^{k} \beta_{jj} x_j^2 \tag{6}$$

where β_{jj}'s represent the quadratic term coefficients.

The addition of more coefficients to the model implies the necessity of a higher number of runs to allow a reliable coefficient estimation. This increase is usually done by adding axial points to the factorial design, resulting in the central composite design (CCD). The impact of the axial point addition can be visualized in **Figure 2**, where the study space is larger when compared to the factorial design (**Figure 1**).

The codified α level depends on the number of factors:

$$\alpha = 2^{k/4} \tag{7}$$

For 2 and 3 factors, α becomes equal to 1.41 and 1.68; respectively. The usual treatment combinations are presented in **Table 3**.

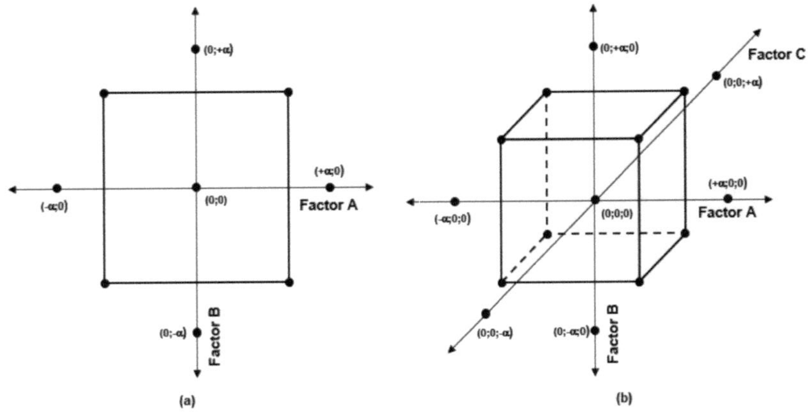

Figure 2. Geometric view of central composite designs: (a) two factors and (b) three factors.

(a)

Run	Factor A	Factor B
1	−1	−1
2	1	−1
3	−1	1
4	1	1
5	−1.41	0
6	+1.41	0
7	0	−1.41
8	0	+1.41
9	0	0
10	0	0
11	0	0

(b)

Run	Factor A	Factor B	Factor C
1	−1	−1	−1
2	1	−1	−1
3	−1	1	−1
4	1	1	−1
5	−1	−1	1
6	1	−1	1
7	−1	1	1
8	1	1	1
9	−1.68	0	0
10	+1.68	0	0
11	0	−1.68	0
12	0	+1.68	0
13	0	0	−1.68
14	0	0	+1.68
15	0	0	0
16	0	0	0
17	0	0	0

Table 3. Central composite designs: (a) two factors; (b) three factors.

In **Table 3**, runs 5–8 (for 2 factors, **Table 3a**) and 9–14 (for 3 factors, **Table 3b**) are called axial points. For the cases discussed above, Eq. (6) becomes:

$$\boldsymbol{k = 2factors}: y = \beta_0 + \beta_1 x_1 + \beta_2 x_2 + \beta_{12} x_1 x_2 + \beta_{11} x_1^2 + \beta_{22} x_2^2 \tag{8}$$

$$\boldsymbol{k = 3factors}: y = \beta_0 + \beta_1 x_1 + \beta_2 x_2 + \beta_3 x_3 + \beta_{12} x_1 x_2 + \beta_{13} x_1 x_3 + \beta_{23} x_2 x_3 + \beta_{11} x_1^2 + \beta_{22} x_2^2 + \beta_{33} x_3^2 \tag{9}$$

The statistical analysis of central composite design follows the steps presented for the 2^k factorial design, now considering the quadratic terms.

3.3. Box-Behnken design (BBD)

Box-Benhken is a three-level DoE that comprises 2^k factorial points with incomplete block design and is used to fit second-order model regression. For three factors, this methodology corresponds to spherical revolving design, corresponding to a centre point surrounded by the middle points of the edges of a cube circumscribed on a sphere of radius $\sqrt{2}$ [27] (**Figure 3**).

Figure 3 shows that Box-Behnken design do not contain any points at the vertices of the cubic region delimitated by the upper and lower levels of each factor (corner points), making this

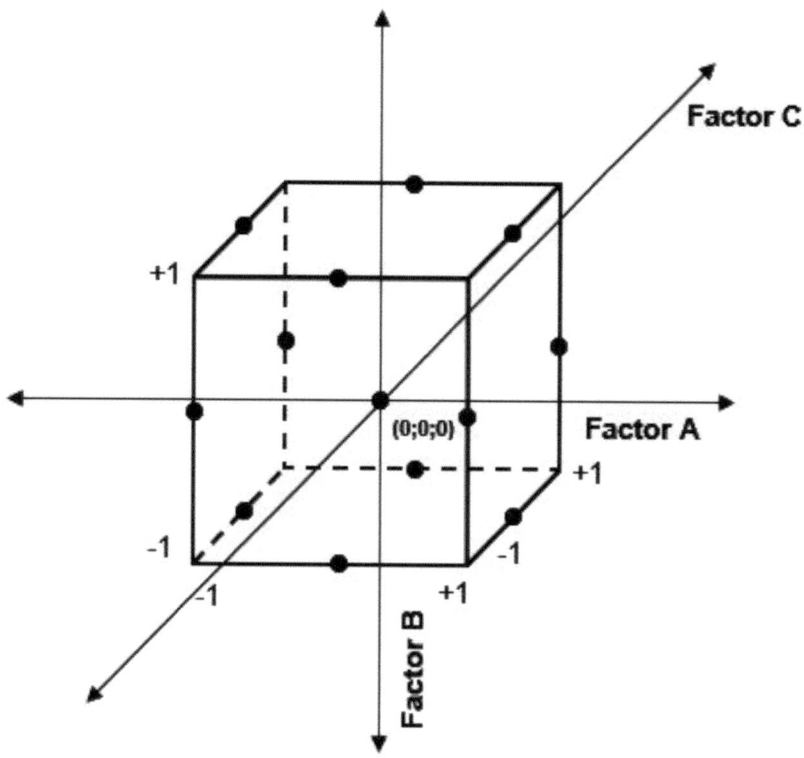

Figure 3. Geometric view of a three-factor Box-Behnken design.

methodology advantageous when these points represent expensive or physically impossible experimental conditions [27]. **Table 4** represents a three factor Box-Behnken design matrix.

3.4. Optimization

Box-Behnken design is also an efficient response surface methodology, since it requires a lower number of runs when compared to central composite design, as can be seen comparing **Tables 3** and **4**.

When a design of experiments is performed, the goal resides in obtaining factor values that optimize (maximize or minimize) the response variables. In general, this objective could be achieved using a sequential strategy, in which a factorial design (2^k FD, for example) followed by a response surface methodology (CCD, for example) are employed in sequence. This procedure is important, since it allows to fit (correct) the factors range of variation and remove non-significant factors from the study.

However, a problem arises when multiple responses are evaluated and need to be optimized simultaneously. This optimization could be performed using several approaches such as to overlay the response surfaces (which is useful for less than three factors); formulation and resolution of constrained optimization problems, using nonlinear programming methods and the desirability functions.

	Factor		
Run	**A**	**B**	**C**
1	−1	−1	0
2	−1	+1	0
3	+1	−1	0
4	+1	+1	0
5	−1	0	−1
6	−1	0	+1
7	+1	0	−1
8	+1	0	+1
9	0	−1	−1
10	0	−1	+1
11	0	+1	−1
12	0	+1	+1
13	0	0	0
14	0	0	0
15	0	0	0

Table 4. Three factor Box-Behnken design.

http://dx.doi.org/10.5772/68097

The use of desirability functions consists in converting all the m response variables (y_i) into individual desirability functions (d_i), making them vary from 0 to 1, which means non-achievement and achievement of the goal, respectively. Thus, the overall desirability function (D) is built according to Eq. (10), and the individual desirability functions are varied in order to optimize D.

$$D = (d_1.d_2...d_m)^{1/m} \tag{10}$$

4. Advanced oxidation processes: photo-Fenton technology

Many organic chemicals discharged into the aquatic environment are not only toxic but also only partly biodegradable and are not easily degraded by conventional biological wastewater treatment plants. That is the reason to develop effective methods aiming degradation of chemical pollutants, either to less noxious transformation products or to their complete mineralization (mainly to CO_2 and H_2O molecules). Throughout the last decades, new methods for water and wastewater cleaning processes, called as advanced oxidation technologies, have received more attention. High rates of pollutant oxidation, flexibility concerning water quality variations and small dimensions of reactors are some advantages of such processes. On the other hand, high operating costs and special safety requirements because of the use of very reactive chemicals (ozone, hydrogen peroxide, etc.) and high-energy sources (UV lamps, electron beams, radioactive sources) are the main negative concerns about AOT.

Advanced oxidation technologies (AOT) imply the use of powerful oxidizing intermediates (the hydroxyl radical OH) which can oxidize and degrade primarily organic pollutants from air and water. The term advanced is used because the chemical reactions involved are essentially the same (except billions of times faster) as the reactions that would occur if these pollutants were exposed in a natural environment. The ubiquitous occurrence of hydroxyl radicals ($^{\bullet}OH$) in various types of environments that include natural waters, the atmosphere, biological systems and interstellar space is now well established. Hydroxyl radicals were first discovered in 1934 by Haber and Weiss in what is known today as the Fenton reaction [28]. It is now well known that, under most atmospheric conditions, $^{\bullet}OH$ radicals govern the oxidative capacity of the natural atmosphere. $^{\bullet}OH$ radicals are composed of a hydrogen atom bonded to an oxygen atom which makes them highly reactive, readily stealing hydrogen atoms from other molecules to form water molecules [29].

AOTs have been defined by Glaze et al. [1] as 'near ambient temperature and pressure water treatment processes which involve the generation of a very powerful oxidizing agent such as hydroxyl radical ($^{\bullet}OH$) in solution in sufficient quantity to effective water purification'. AOTs are applied whenever conventional oxidation techniques are insufficient, when process kinetics becomes very slow, or because contaminants are refractory to chemical oxidation in aqueous medium or partially oxidized yielding stable transformation products showing even greater toxicity than the starting pollutants. However, it must also be taken into consideration that the oxidation ability of most AOT diminishes considerably when treating high organic matter contents (>5.0 g L^{-1}), thereby requiring the consumption of excessive amounts of expensive reactants that makes the treatment far less cost-affordable.

AOTs oxidize a broad range of contaminants, including those that are not readily removed with other advanced technologies (e.g. reverse osmosis or granular activated carbon). Most of the commercially viable AOT use either ozone or photochemical processes [i.e. ultraviolet (UV) or visible light] to generate OH [30]. In AOT, $^{\bullet}OH$ radicals are generated usually by coupled chemical and/or physical systems that include $H_2O_2/Fe(II)$ or $H_2O_2/Fe(III)$ (Fenton), $H_2O_2/$ catalyst or peroxide/catalyst (Fenton-like), O_3 (ozonation) and H_2O_2/O_3 (peroxone) that are often associated with an irradiation technique, namely vacuum-UV radiation, UV radiation (low-, medium-, or high-pressure lamps), pulse radiolysis or ultrasound [29].

Since its first use [28], until now, the most effective and simple way of generating hydroxyl radicals for organic pollutants degradation is by classical Fenton reaction (Eq. (11)) [31].

$$Fe^{2+} + H_2O_2 \rightarrow Fe^{3+} + \cdot OH + OH^- \ (k = 74\ M^{-1}\ s^{-1}) \tag{11}$$

However, one of the main inconveniences of Fenton reaction is the rapid consumption of Fe^{2+} (Eq. (11)) and a very slow regeneration of Fe^{2+} by 'Fenton-like' reaction (Eq. (12)). Fenton-based degradation process, therefore, demands a high initial concentration or a continuous dosage of Fe^{2+}. This disadvantage is overcome in photo-Fenton process wherein (UV-Visible irradiation, $\lambda < 550$ nm), Fenton/Fenton-like reaction (Eq. (11)) is paired with UV irradiation to regenerate Fe^{2+} (Eq. (13)), thereby minimizing the Fe^{2+} dosage for degradation [31].

$$Fe^{3+} + H_2O_2 \rightarrow Fe^{2+} + HO_2^{\cdot} + H^+ \ (k' = 0.02\ M^{-1}s^{-1}) \tag{12}$$

$$Fe^{3+} + H_2O_2 \xrightarrow{hv} Fe^{2+} + \cdot OH + H^+ \tag{13}$$

Consequently, the photo-Fenton has been recognized as a prospective UV-based AOT for treatment of industrial waste water. In photo-Fenton process, Fe(III) aqua species, mainly the hydroxo complex $[Fe(H_2O)5OH]^{2+}$ formed at pH~3 is photoactive and regenerates Fe(II) upon UV irradiation [31]. In the photo-Fenton process, only iron is catalytic, whereas hydrogen peroxide plays a sacrificial role. The mechanism of the process is very complex and still remains incompletely elucidated; it is widely accepted that hydroxyl radical plays a major role as oxidizing agent, although the involvement of other species, such as high-valence iron, has not been ruled out [32].

When it comes to industrial applications, the scientific literature is scarce and, basically, in the same situation observed by Vogelpohl [33]. There is a lack of published data providing comparisons with bench-scale or pilot-scale data. Moreover, studies developed by companies are often only disclosed internally. As noted by Vogelpohl [33], there are few data on costs, both for installation and for operation, without which there will be a considerable gap between academia and industry. On the other hand, the growing number of patents in the area is a positive factor that indicates that the acceptance of AOTs by the industrial sector is growing each year. By performing a simple search (5 January 2017), through the Google Patents robot, the expression 'advanced oxidation process*' generated a total of 3384 results, the keyword 'fenton' generated 22,977 results, whereas 'wastewater AND pharmaceut* AND fenton' strategy obtained 1379 results. In this bibliographic research, other forms of publication other than patents were excluded.

Despite scarce literature information—compared to academic number of papers—about industrial applications of AOT for water and air remediation, advanced oxidation processes have been successfully commercialized during recent years [34]. Basically, advanced oxidation solutions begin with a wastewater chemical study and a geologic data evaluation (for treatment of contaminated soil and groundwater). A critical step in ensuring the success of a treatment program is the selection of the appropriate chemical reagent and application method. An initial bench-scale is useful to select the appropriate method. Previous information from the scientific literature (reports, academic studies, conference proceedings and papers) is mandatory to the chemical remediation industry. In commercial scale of Fenton's reagent, H_2O_2 dosage of 5–35% (w/w) is applied. The initial weight of H_2O_2 and Fe^{2+} are based on contaminants levels, chemical characteristics, type of soil to be treated, and the specific ratio of H_2O_2:Fe^{2+} determined during the laboratory study (bench scale previous tests). Occasionally, additional reagents may be applied because of heterogeneity of the medium and to diminish H_2O_2 rate of decomposition aiming additional contact time for the contaminants. If the natural pH of the contaminant site is not low enough for efficient hydroxyl radical generation, H_2SO_4 may be added to adjust pH prior to the Fenton's reagent application. Some commercial applications of AOT are shown as follows:

Calgon Carbon Corporation has provided advanced oxidation technologies for disinfection drinking water, municipal wastewater and industrial wastewater with low-pressure and medium-pressure lamps UV reactors (SENTINEL®, C3 Series™, and RAYOX® systems). According to Calgon Carbon Corp., each Sentinel® reactor is able to treat 200 million litres of industrial effluents a day, accommodating pipe size from 12 to 48 inches. On the other hand, Rayox® has reactors from lab scale (batch test units with 1 kW, 30 kW, 60 kW and 90 kW) to large applications with patented lamp-cleaning device.

The Geo-Cleanse® process is a patented, *in-situ* chemical oxidation technology that utilizes Fenton's reagent and modified-Fenton's reagent to destroy organic compounds from wastewater and soil. As an example of AOT application, Geo-Cleanse International treated groundwater contaminated with high concentrations of perchloroethylene (PCE) and its transformation products at a landfill (25-acre area) located on the Naval Submarine Base, Kings Bay Georgia. The source of PCE contamination was identified on the perimeter of the landfill with concentrations of over 9000 $\mu g\ L^{-1}$. Injection of chemical reagents and catalyst was implemented by the placement of 23 injection wells. The AOT achieved over a 98% destruction of chlorinated hydrocarbons.

Another *in-situ* chemical oxidation solution is provided by CleanOX® that is based on the Fenton's reagent. In-Situ Oxidative Technologies Inc. provides ISOTEC's modified Fenton's reagent (MFR), which consists of injecting patented chelated iron catalysts and stabilized hydrogen peroxide into contaminated aquifers (at pH~7).

In-Situ Technieken (The Netherlands) treats contaminated soil also with Fenton's reagent with a process that can be combined (or not) with aerobic/anaerobic biological degradation (BISCO®). The Dow Chemical produces raw materials for the polyurethane industry in Delfzijl (The Netherlands). The soil at the site of the sandblasting area was polluted with monochlorobenzene (MCB). The DOW plant in Delfzijl was obliged to submit the principles for a soil remediation plan to the Provincial Government of Groningen. The remediation was carried out with AOT and in accordance with the Netherlands Directive for Soil Protection (NRB).

On-Contact Remediation Process® by Environmental Business Solutions International Inc. (EBSI) is another commercial AOT process. In one of the examples provided by EBSI Inc., *in-situ* remediation was successfully implemented in Boston, Massachusetts. Fuel oil, gasoline and several plasticizer additives, including bis-2-ethylhexyl phthalate (DEHP) were stored in underground storage tanks (UST) outside the facility loading bays. Light Non-Aqueous Phase Liquid, consisting of fuel oil and DEHP, was discovered at the site in the late 1980s and various methods to remove it had been attempted over many years. After three rounds of oxidizer treatment, LNAPL levels were reduced by more than 70% by AOT.

There has been a tendency to quote treatment costs per unit volume for a waste stream and technology (e.g. dollars/1000 gal); however, such notation does not consider the concentration of the contaminant nor the treatment goals. Bolton et al. [35] proposed figures-of-merit that are based on electrical energy consumption within two phenomenological kinetic order regimes: one for high contaminant concentrations and one for low concentrations. A simple understanding of the overall kinetic behaviour of organic destruction in a waste stream (i.e. whether zero or first order) is necessary for describing meaningful electrical efficiencies. These standard figures-of-merit are valuable in that they give a direct link to the electrical efficiency of an advanced oxidation process, independent of the nature of the system and therefore also allow comparison of widely disparate AOT. Such figures-of-merit are necessary not only to compare AOT, but also to provide the requisite data for scale-up and economic analyses for comparison with conventional treatment technologies (e.g. carbon adsorption/regeneration, air stripping and incineration).

There are many important factors in selecting a waste-stream treatment technology including economics, economy of scale, regulations, effluent quality goals, operation (maintenance, control and safety) and robustness (flexibility to change/upsets). Although all these factors are important, economics is often sovereign. A full economic analysis of the net present cost (i.e. amortized investment, installation and operating costs) of implementing a wide range of treatment technologies represents an arduous task and often can be both site- and problem-specific. A simple figure-of-merit based on electrical energy consumption can be very useful and informative for AOT, since these processes are often electrical energy intensive and electrical energy can represent a major fraction of the operating costs. Moreover, electrical energy dosage requirements also dictate the size of the capital equipment needed to generate the requisite dosage, thus investment should also tend to scale with this figure-of-merit [35].

Asaithambi et al. [36] compared the performance of the photo (UV), Fenton, photo-Fenton and ozone-photo-Fenton processes in terms of colour removal and chemical oxygen demand (COD) removal of distillery industrial effluent together with the associated electrical energy per order. The colour and COD removals of industrial effluents were investigated using the pseudo first-order kinetic model. It was observed from the experimental results that $O_3/UV/Fe^{2+}/H_2O_2$ process yielded a 100% colour and 95.50% COD removals with electrical energy per order of 0.015 $kWhm^{-3}$ compared to all other combinations of AOT.

Lin et al. [37] investigated the degradation of ofloxacin antibiotics by UV/H_2O_2 process in a large photoreactor, and the effects of UV wavelength, H_2O_2 dosage and pH on this process. The degradation of ofloxacin proceeded more rapidly under UV-254 nm than under UV-365 nm. The degradation of ofloxacin by the UV/H_2O_2 process followed pseudo-first-order kinetics. At pH 3 and an H_2O_2 dosage of 0.27 g L^{-1}, 97% of ofloxacin was degraded under UV-254 nm after 30 min. The electrical energy per order of removal (EE/O) figure-of-merit value for the treatment of 10 mg L^{-1} ofloxacin by the UV-365 nm/H_2O_2 process was 22.5 kWh m^{-3} per order. However, that of the UV-254 nm/H_2O_2 process was significantly lower, at 2.2 kWh m^{-3} per order.

Lin and Wu [38] studied the effectiveness of the UV/$S_2O_8^{2-}$ process in the degradation of ciprofloxacin antibiotics in aqueous solutions without adjusting their pH using a large photoreactor. The EE/O values for 10 mg L^{-1} ciprofloxacin treatment by the UV/$S_2O_8^{2-}$ process at various $Na_2S_2O_8$ concentrations were calculated. The lowest EE/O value in the UV/$S_2O_8^{2-}$ process was calculated as 0.653 kWh m^{-3} per order at $Na_2S_2O_8$ concentration of 3.84 g L^{-1}.

Only a small fraction of the scientific papers related to advanced oxidative technologies makes use of the electrical energy consumption figure-of-merit. In our understanding, one of the best response variables that could be used in the study of wastewater degradation (pharmaceuticals or not) would be EE/O. With EE/O as variable response in a design of experiments, not only the main factors associated to the advanced oxidative processes (light source, catalyst, H_2O_2, O_3, Fe^{2+}, etc.), but also the kinetics of the reaction and the energy cost would be used in an experimental study to reach the objective of the process. There is a gap in the scientific community regarding the use of the EE/O tool that could be explored by researchers in conjunction with the traditional statistical tool design of experiments.

Acknowledgements

This work was funded in part by Federal Institute of Sao Paulo (PRP-IFSP #226/2016).

Author details

André Luís de Castro Peixoto[1]*, Ademir Geraldo Cavallari Costalonga[1],
Mateus Nordi Esperança[1] and Rodrigo Fernando dos Santos Salazar[2]

*Address all correspondence to: alcpeixoto@ifsp.edu.br

1 Grupo de Química Tecnológica (GQT), Federal Institute of Sao Paulo (IFSP), Campus Capivari, Capivari, Brazil

2 Centro de Ciências da Saúde e Agrárias, University of Cruz Alta (UNICRUZ), Campus Dr. Ulysses Guimarães, Cruz Alta, Brazil

References

[1] Glaze WH, Kang JW, Chapin DH. The chemistry of water treatment processes involving ozone, hydrogen peroxide and ultraviolet radiation. Ozone: Science & Engineering. 1987;**9**(4):335-352. DOI: 10.1080/01919518708552148

[2] Parsons S, editor. Advanced Oxidation Processes for Water and Wastewater Treatment. IWA publishing; London, 2004

[3] Araújo KS, Antonelli R, Gaydeczka B, Granato AC, Malpass GRP. Advanced oxidation processes: A review regarding the fundamentals and applications in wastewater treatment and industrial wastewater. Ambiente e Agua-An Interdisciplinary Journal of Applied Science. 2016;**11**(2):387-401. DOI: 10.4136/ambi-agua.1862

[4] Nogueira RFP, Trovó AG, da Silva MRA, Villa RD, de Oliveira MC. Fundamentos e aplicações ambientais dos processos Fenton e foto-Fenton. Química Nova. 2007;**30**(2):400-408. DOI: 10.1590/S0100-40422007000200030

[5] Qu X, Alvarez PJ, Li Q. Applications of nanotechnology in water and wastewater treatment. Water Research. 2013;**47**(12):3931-3946. DOI: 10.1016/j.watres.2012.09.058

[6] Sa RM, Premalatha M. Applications of nanotechnology in wastewater treatment: A review. Imperial Journal of Interdisciplinary Research. 2016;**2**(11):1500-1511

[7] Brown SD. The chemometrics revolution re-examined. Journal of Chemometrics. 2017;**31**(1):1-23. DOI: 10.1002/cem.2856

[8] Myers RH, Montgomery DC, Anderson-Cook CM. Response Surface Methodology: Process and Product Optimization using Designed Experiments. 3rd ed. New Jersey: John Wiley & Sons; 2016. p. 704

[9] Ay F, Kargi F. Advanced oxidation of amoxicillin by Fenton's reagent treatment. Journal of Hazardous Materials. 2010;**179**:622-627. DOI: 10.1016/j.jhazmat.2010.03.048

[10] Ay F, Kargi F. Effects of reagent concentrations on advanced oxidation of amoxicillin by Photo-Fenton treatment. Journal of Chemical Engineering. 2011;**137**(6):472–480. DOI: 10.1061/(ASCE)EE.1943-7870.0000344

[11] Irani M, Rad RR, Pourahmad H, Haririan I. Optimization of the combined adsorption/photo-Fenton method for the simultaneous removal of phenol and paracetamol in a binary system. Microporous and Mesoporous Materials. 2015;**206**:1-7. DOI: 10.1016/j.micromeso.2014.12.009

[12] Diniz LM. Avaliação do reagente de fenton e foto-fenton na remoção de matéria orgânica e toxicidade em um efluente hospitalar [thesis]. Belo Horizonte: Federal University of Minas Gerais; 2015

[13] Marcelino RBP. Aplicação de processos oxidativos avançados para o tratamento de efluente da produção de antibióticos [thesis]. Belo Horizonte: Federal University of Minas Gerais; 2014

[14] Dwivedi K, Morone A, Chakrabarti T, Pandey RA. Evaluation and optimization of Fenton pretreatment integrated with granulated activated carbon (GAC) filtration for carbamazepine removal from complex wastewater of pharmaceutical industry. Journal of Environmental Chemical Engineering. 2016, DOI: 10.1016/j.jece.2016.12.054

[15] Silva VV. Degradação de amoxicilina por Fenton e foto-Fenton [thesis]. Porto Alegre: Federal University of Rio Grande do Sul; 2015

[16] Pérez-Moya M, Graells M, Castells G, Amigó J, Ortega E, Buhigas G, Pérez LM, Mansilla HD. Characterization of the degradation performance of the sulfamethazine antibiotic by photo-Fenton process. Water Research. 2010;**44**:2533-2540. DOI: 10.1016/j.watres.2010.01.032

[17] Silva SS, Chiavone-Filho O, Barros-Neto EL, Foletto EL, Mota ALN. Effect of inorganic salt mixtures on phenol mineralization by Photo-Fenton-Analysis via an experimental design. Water Air and Soil Pollution. 2014;**225**:1784. DOI: 10.1007/s11270-013-1784-x

[18] Affam AC, Chaudhuri M, Kutty SRM. Optimization of modified Fenton (FeGAC/H2O2) pretreatment of antibiotics. Pertanika Journal of Science & Technology. 2014;**22**(1):239-254

[19] Almeida LC. Otimização de processo de mineralização de compostos orgânicos utilizando sistemas eletro-fenton e fotoeletro-fenton por irradiação UV artificial e solar [thesis]. São Carlos: Federal University of São Carlos; 2011

[20] Frade VMF. Oxidação química de enrofloxacina pelo processo Fenton [thesis]. São Paulo: University of São Paulo; 2013

[21] Homem V, Alves A, Santos L. Amoxicillin degradation at ppb levels by Fenton's oxidation using design of experiments. Science of the Total Environment. 2010;**408**:6272-6280. DOI: 10.1016/j.scitotenv.2010.08.058

[22] Rozas O, Contreras D, Mondaca MA, Pérez-Moya M, Mansilla HD. Experimental design of Fenton and photo-Fenton reactions for the treatment of ampicillin solutions. Journal of Hazardous Materials. 2010;**177**:1025-1030. DOI: 10.1016/j.jhazmat.2010.01.023

[23] Sarrai AE, Hanini S, Merzouk K, Tassalit D, Szabó T, Hernádi K, Nagy L. Using central composite experimental design to optimize the degradation of Tylosin from aqueous solution by Photo-Fenton reaction. Materials. 2016;**428**(9):428. DOI: 10.3390/ma9060428

[24] Zaidan LEMC, Silva AMRB, Sales RVL, Salgado JBA, Moraes SCG, Souza DP, Galvão CC, Rodríguez-Díaz JM, Napoleão DC, Benachour M, Silva VL. Optimization of phenol degradation and its derivatives using photo-Fenton and application industrial. Chemical and Process Engineering Research. 2016;**42**:44-52

[25] Arslan-Alaton I, Ayten N, Olmez-Hanci T. Photo-Fenton-like treatment of the commercially important H-acid: Process optimization by factorial design and effects of photocatalytic treatment on activated sludge inhibition. Applied Catalysis B: Environmental. 2010;**96**:208-217. DOI: 10.1016/j.apcatb.2010.02.023

[26] Domínguez JR, González T, Palo P, Cuerda-Correa EM. Fenton + Fenton-like integrated process for carbamazepine degradation: Optimizing the system. Industrial & Engineering Chemistry Research. 2012;**51**:2531-2538. DOI: 10.1021/ie201980p

[27] Montgomery DC. Design and Analysis of Experiments. 6th ed. John Wiley & Sons Inc; New Jersey, 2005

[28] Fenton HJH. LXXIII—Oxidation of tartaric acid in presence of iron. Journal of the Chemical Society, Transactions. 1894;**65**:899-910. DOI: 10.1039/CT8946500899

[29] Gligorovski S, Strekowski R, Barbati S, Vione D. Environmental implications of hydroxyl radicals (•OH). Chemical Reviews. 2015;**115**(24):13051-13092. DOI: 10.1021/cr500310b

[30] Collins J, Bolton J. Advanced Oxidation Handbook. 1st ed. Denver: American Water Works Association; 2016;1-194

[31] Subramanian G, Madras G. Introducing saccharic acid as an efficient iron chelate to enhance photo-Fenton degradation of organic contaminants. Water Research. 2016;**104**:168-177. DOI: 10.1016/j.watres.2016.07.070

[32] Marin ML, Santos-juanes L, Arques A, Amat AM, Miranda MA. Organic photocatalysts for the oxidation of pollutants and model compounds. Chemical Reviews. 2012;**112**:1710-1750. DOI: 10.1021/cr2000543

[33] Vogelpohl A. Applications of AOPs in waste water treatment. Water Science and Technology. 2007;**55**(12):207-211

[34] Oppenländer T. Photochemical Purification of Water and Air. 1st ed. Weinheim: Wiley-VCH; 2003. p. 368

[35] Bolton JR, Bircher KG, Tumas W, Tolman CA. Figures of merit for the technical development and application of advanced oxidation technologies for both electric and solar driven systems. 2001;**73**(4):627-637 http://dx.doi.org/10.1351/pac200173040627

[36] Asaithambi P, Sajjadi B, Raman A, Aziz A. Integrated ozone-photo-Fenton process for the removal of pollutant from industrial wastewater. Chinese Journal of Chemical Engineering. 2016, DOI: 10.1016/j.cjche.2016.10.005

[37] Lin CC, Lin HY, Hsu LJ. Degradation of ofloxacin using UV/H2O2 process in a large photoreactor. Separation and Purification Technology. 2016;**168**:57-61. DOI: 10.1016/j.seppur.2016.04.052

[38] Lin CC, Wu MS. Degradation of ciprofloxacin by UV/$S_2O_8^{2-}$ process in a large photoreactor. Journal of Photochemistry and Photobiology A: Chemistry. 2014;**285**:1-6. DOI: 10.1016/j.jphotochem.2014.04.002

Chapter 4

Model-Based Evolutionary Operation Design for Batch and Fed-Batch Antibiotic Production Bioprocesses

Samuel Conceição de Oliveira

Additional information is available at the end of the chapter

http://dx.doi.org/10.5772/intechopen.69395

Abstract

The control policy determination for batch and fed-batch antibiotic production bioprocesses is an important practical issue due to the high added value of these bioproducts. Since it is highly desirable to optimize the antibiotic production, several methods have been proposed aimed at this objective. Once having a mathematical model for the bioprocess, the optimization problem can be formulated within the framework of Pontryagin's maximum principle and of the optimal control theory to determine the best control trajectory for certain key manipulated variables, such as temperature, pH, and substrate feed rate. In this chapter, applications of these model-based techniques to optimize and control antibiotics production bioprocesses are reviewed and new aspects are emphasized. The cases analyzed included the optimization of the substrate feed rate in a fed-batch reactor and of the temperature in a batch reactor during penicillin fermentations. The main contributions of this study were: (i) the proposition of a different procedure for calculating the second switching time of substrate feed rate, (ii) the application of simpler numerical methods to solve the two-point boundary-value problem associated with the temperature profile optimization, and (iii) the demonstration that the non-isothermal operation is more productive in antibiotic than the operation under constant temperature.

Keywords: modeling, optimization, evolutionary operation, bioprocesses, antibiotic fermentation, batch bioreactors, fed-batch bioreactors

1. Introduction

Improvement in the productivity of many submerged fermentation processes is carried out by manipulating nutritional and physical parameters such as medium composition, agitation speed, aeration rate, pH, and temperature [1, 2]. Although the attainment of optimal conditions for a multivariable fermentation process is often tedious, it is possible to undertake a rational procedure by using statistical experimental designs [2].

Experimental designs can be divided in two distinct groups [3–6]: (i) model-based experimental designs and (ii) statistical experimental designs. In model-based experimental designs, predictions of a mathematical model are used to determine how an experiment or process should be performed, whereas, with statistical experimental designs, these model predictions are not explicitly required.

The optimization and operation of fermentation processes play a key role in the biotechnology industry due to heavy competition among companies. Secondary metabolites, such as antibiotics and other pharmaceutical products, represent an important added value; therefore, improvements in the production of these bioproducts are of great interest to industries. To achieve high-performance operations, the optimization of manipulated variables that affect the fermentation process becomes a significant task.

In general, optimization problems can be classified in two categories: set-point and profile optimizations [7]. Set-point optimization problems involve finding the best set of values of manipulated variables that lead to the maximization of performance indexes [7]. Profile optimization consists of determining temporal or spatial functions (profiles), rather than a point in n-dimensional space, which lend an optimal value to the performance index [7].

In antibiotic fermentation, it is well known that the temperature and pH for the maximum rate of antibiotic production are different from those for the maximum rate of cell growth [8]. In this sense, the implementation of temperature and pH profiles plays an important role in significant improvements in antibiotic production bioprocesses [8].

Since the primary goal of a fermentation process is the cost-effective production of bioproducts, it is important to select the more appropriate operating mode that allows the production of the desired product at a high concentration with a high productivity and yield [9]. Fed-batch bioprocesses have been widely employed for the production of various bioproducts, including primary and secondary metabolites [9]. In the particular case of secondary metabolites, such as antibiotics, the interaction between growth metabolism and product biosynthesis is critically affected by growth-limiting nutrient concentrations. Since both the underfeeding and the overfeeding of nutrients are detrimental to cell growth and product formation, due to the occurrence of phenomena such as cell starvation and catabolite repression, establishing a suitable feeding strategy is crucial in fed-batch bioprocesses [9, 10].

A particular time sequence of control variables may be required in order to conduct the bioprocess over time in a trajectory that provides the greatest productivity. This can lead to complex optimal time profiles for the control variables, which are sometimes impossible to be determined purely experimentally. Thus, appropriate mathematical and numerical methods can be applied for the determination of these profiles in order to reduce the experimental effort and the required time for optimization.

The search for the optimal pH, temperature, and substrate feed-rate profiles in batch and fed-batch antibiotic fermentation is a typical problem of optimization and evolutionary operations for which the use of kinetic models and powerful mathematical techniques is essential for their solution [7, 10, 11]. According to Rani and Rao [12], several approaches for the determination of optimal time profiles for control variables have been reported in the literature [13–15]. In these reports, the optimization problem is generally formulated on the basis of Pontryagin's maximum principle, taking as a starting point a phenomenological mathematical model of the bioprocess. For simple mathematical models, the problem can be solved analytically, from the Hamiltonian of the system, by applying an iterative scheme on the control variable to determine the optimal control profile [8, 16–19].

In this chapter, two studies on the optimization and the evolutionary operation of antibiotic production bioprocesses are revisited, and new results are obtained and highlighted. Such studies report mathematical models of bioprocesses, in conjunction with Pontryagin's maximum principle, to optimize the substrate feed-rate profile for a fed-batch bioreactor and the temperature profile for a batch fermentation in order to maximize the production of antibiotic. The fundamentals of Pontryagin's maximum principle, when applied to the cases analyzed, are also presented.

The aim is to provide a theoretical basis for the application of a model-based methodology that can be used for the optimization and control of bioprocesses from other antibiotics and secondary metabolites with a broad structural diversity and therapeutic activity, including antibacterial, antifungal, antiviral, antitumor, immunosuppressive, antihypertensive, and antihypercholesterolemic compounds.

2. Case studies: batch and fed-batch antibiotic production bioprocesses

During batch and fed-batch bioprocesses, the state variables (cell, substrate, oxygen and product concentrations, temperature, and pH) change significantly, from initial to final values. This dynamic behavior motivates the development of optimization methods to find the optimal time trajectories for the control variables in order to improve the performance of these bioprocesses.

Two case studies on the optimization of control variables in batch and fed-batch antibiotic production bioprocesses are revisited, and additional results are obtained and presented. The cases studied are those reported by Costa [20] and Constantinides and Mostouffi [17], concerning the optimization of the substrate feed rate in a fed-batch reactor and the temperature in a batch reactor, respectively. These cases are presented and detailed in the following sections.

2.1. Case study #1: determination of the optimal substrate feed-rate profile in fed-batch bioreactor for penicillin production

In this case study, the bioprocess of penicillin production by *Penicillium chrysogenum* is described by the mathematical model presented by Costa [20], which is based on the classical model proposed by Bajpai-Reuss for penicillin fermentation. In this model, the specific growth rate (μ) takes into account diffusional limitations that occur in the filamentous fungal biomass, as described by the Contois model. The specific rate of product formation (π) considers that penicillin production is repressed by high substrate concentrations (catabolic repression), being modeled by the Andrews equation. Penicillin degradation by hydrolysis is also considered, assuming first-order kinetics for this reaction. The specific rate of substrate consumption (σ) is represented by the Herbert-Pirt generalized model, whereby the substrate is consumed for cell growth and maintenance and for product formation. The equations of the full mathematical model are as follows (note that if the dilution rate is used as the control variable, the total mass-balance equation is not required for the optimization problem formulation, thus reducing the equation system dimension):

$$\frac{dX}{dt} = \mu X - Xu; \; \mu = \frac{\mu_m S}{BX + S} \tag{1 – 2}$$

$$\frac{dS}{dt} = -\sigma X + (S_f - S)u; \; \sigma = \frac{\mu}{Y_{X/S}} + \frac{\pi}{Y_{P/S}} + m \tag{3 – 4}$$

$$\frac{dP}{dt} = (\pi X - k_h P) - Pu; \; \pi = \frac{\pi_m S}{k_m + S + \frac{S^2}{k_i}} \tag{5 – 6}$$

where

- X, S, and P denote the concentrations of cell, substrate, and product, respectively;
- μ, σ, and π are the specific rates of cell growth, substrate consumption, and product formation;
- μ_m, B, $Y_{X/S}$, $Y_{P/S}$, m, π_m, k_m, and k_i are the parameters of the mathematical model, including kinetic and yield parameters;
- $u = D = F/V : u =$ control variable; $D =$ dilution rate; $F =$ feed rate; $V =$ culture volume (7)

In matrix notation:

$$\frac{d\underline{X}}{dt} = \underline{f}(\underline{X}) + \underline{g}(\underline{X})\, u \tag{8}$$

where

$$\underline{X} = \begin{bmatrix} X \\ S \\ P \end{bmatrix}; \; \underline{f}(\underline{X}) = \begin{bmatrix} \mu X \\ -\sigma X \\ \pi X - k_h P \end{bmatrix}; \; \underline{g}(\underline{X}) = \begin{bmatrix} -X \\ (S_f - S) \\ -P \end{bmatrix} \tag{9 – 11}$$

The constraints imposed on the control variable u are as follows:

$$u_{\min} \leq u \leq u_{\max} \tag{12}$$

where $u_{\min} = 0$ and $u_{\max} = F_{\max} / V$.

Another constraint concerns the maximum volume of culture (final volume), i.e., $V(t_f) = V_{max} = V_f$, where t_f is the final processing time.

The initial conditions are given by

$$X(0) = X_0;\ S(0) = S_0;\ P(0) = P_0;\ V(0) = V_0 \tag{13}$$

The objective of the optimization/control problem is to determine the optimal time profile for the control variable that maximizes the antibiotic concentration at the end of the bioprocess.

According to the Pontryagin's maximum principle, the optimal profile must maximize the Hamiltonian, given by

$$H = \underline{\lambda}^T [\underline{f}(\underline{X}) + \underline{g}(\underline{X})u];\ \underline{\lambda}^T = [\lambda_1\ \ \lambda_2\ \ \lambda_3] \tag{14}$$

or

$$H = H_0(t) + \varphi(t)\ u\ \begin{cases} H_0(t) = \underline{\lambda}^T \underline{f}(\underline{X}) \\ \varphi(t) = \underline{\lambda}^T \underline{g}(\underline{X}) \end{cases} \tag{15}$$

Since $\underline{f}(\underline{X}) = \begin{bmatrix} f_1 \\ f_2 \\ f_3 \end{bmatrix}$ and $\underline{g}(\underline{X}) = \begin{bmatrix} g_1 \\ g_2 \\ g_3 \end{bmatrix}$, the following equations are obtained:

$$H_0(t) = \lambda_1 f_1 + \lambda_2 f_2 + \lambda_3 f_3 = \lambda_1 \mu X - \lambda_2 \sigma X + \lambda_3(\pi X - kP) \tag{16}$$

$$\phi(t) = \lambda_1 g_1 + \lambda_2 g_2 + \lambda_3 g_3 = -\lambda_1 X + \lambda_2(S_f - S) - \lambda_3 P \tag{17}$$

For optimal control, it is established that

- If $\varphi(t) > 0$, then $u = u_{max}$.
- If $\varphi(t) < 0$, then $u = u_{min}$.
- If $\varphi(t) = 0$, then $u = u_{\sin}$.

$$\text{In the singular interval}: H(t) = 0;\ \phi(t) = 0;\ H_0(t) = 0 \tag{18 – 20}$$

Since $\underline{\lambda} = \underline{\lambda}(t)$, the following equations can be developed:

$$\frac{d}{dt}\underline{\lambda} = -\frac{\partial H}{\partial \underline{X}} = -\frac{\partial}{\partial \underline{X}}[\underline{\lambda}^T(f(x) + g(x)u)] = -\underline{\lambda}^T \underbrace{\left(\frac{d}{d\underline{X}}\underline{f}(\underline{X}) + u\frac{d}{d\underline{X}}\underline{g}(\underline{X})\right)}_{\left(\frac{d}{dt}\underline{\lambda}\right)^T} \tag{21}$$

$$\phi(t) = \underline{\lambda}^T \underline{g}(\underline{X}) \Rightarrow \frac{d\phi}{dt} = \left(\frac{d}{dt}\underline{\lambda}\right)^T \underline{g}(\underline{X}) + \underline{\lambda}^T \frac{d}{dt}\underline{g}(\underline{X}) \tag{22}$$

$$\frac{d}{dt}\underline{g}(\underline{X}) = \frac{d}{d\underline{X}}\left(\underline{g}(\underline{X})\right)\frac{d\underline{X}}{dt} = \frac{d}{d\underline{X}}\left(\underline{g}(\underline{X})\right)[\underline{f}(\underline{X}) + \underline{g}(\underline{X})u] \tag{23}$$

$$\frac{d}{dt}\underline{g}(\underline{X}) = \frac{d}{d(\underline{X})}\left(\underline{g}(\underline{X})\right)\underline{f}(\underline{X}) + \frac{d}{d\underline{X}}\left(\underline{g}(\underline{X})\right)\underline{g}(\underline{X})\,u \tag{24}$$

Substituting Eqs. (21) and (24) into Eq. (22) gives

$$\frac{d\varphi}{dt} = -\underline{\lambda}^T\left(\frac{d}{d\underline{X}}\underline{f}(\underline{X}) + u\frac{d}{d\underline{X}}\underline{g}(\underline{X})\right)\underline{g}(\underline{X}) + \underline{\lambda}^T\left(\frac{d}{d\underline{X}}\left(\underline{g}(\underline{X})\right)\underline{f}(\underline{X}) + \frac{d}{d\underline{X}}\left(\underline{g}(\underline{X})\right)\underline{g}(\underline{X})u\right) \tag{25}$$

$$\frac{d\varphi}{dt} = \underline{\lambda}^T\left(\frac{d}{d\underline{X}}\left(\underline{g}(\underline{X})\right)\underline{f}(\underline{X}) - \frac{d}{d\underline{X}}\left(\underline{f}(\underline{X})\right)\underline{g}(\underline{X})\right). \tag{26}$$

By developing the matrices indicated in the previous equation, one obtains

$$\frac{d\varphi}{dt} = [\lambda_1\ \ \lambda_2\ \ \lambda_3] * \left[\begin{bmatrix} \frac{\partial g_1}{\partial X_1} & \frac{\partial g_1}{\partial X_2} & \frac{\partial g_1}{\partial X_3} \\ \frac{\partial g_2}{\partial X_1} & \frac{\partial g_2}{\partial X_2} & \frac{\partial g_2}{\partial X_3} \\ \frac{\partial g_3}{\partial X_1} & \frac{\partial g_3}{\partial X_2} & \frac{\partial g_3}{\partial X_3} \end{bmatrix} * \begin{bmatrix} f_1 \\ f_2 \\ f_3 \end{bmatrix} - \begin{bmatrix} \frac{\partial f_1}{\partial X_1} & \frac{\partial f_1}{\partial X_2} & \frac{\partial f_1}{\partial X_3} \\ \frac{\partial f_2}{\partial X_1} & \frac{\partial f_2}{\partial X_2} & \frac{\partial f_2}{\partial X_3} \\ \frac{\partial f_3}{\partial X_1} & \frac{\partial f_3}{\partial X_2} & \frac{\partial f_3}{\partial X_3} \end{bmatrix} * \begin{bmatrix} g_1 \\ g_2 \\ g_3 \end{bmatrix}\right] \tag{27}$$

where

$$g_1 = -X;\ g_2 = (S_f - S);\ g_3 = -P;\ X_1 = X;\ X_2 = S;\ X_3 = P \tag{28 – 33}$$

$$\frac{\partial g_1}{\partial X_1} = -1;\ \frac{\partial g_1}{\partial X_2} = 0;\ \frac{\partial g_1}{\partial X_3} = 0 \tag{34 – 36}$$

$$\frac{\partial g_2}{\partial X_1} = 0;\ \frac{\partial g_2}{\partial X_2} = -1;\ \frac{\partial g_2}{\partial X_3} = 0 \tag{37 – 39}$$

$$\frac{\partial g_3}{\partial X_1} = 0;\ \frac{\partial g_3}{\partial X_2} = 0;\ \frac{\partial g_3}{\partial X_3} = -1 \tag{40 – 42}$$

$$f_1 = \mu X;\ f_2 = -\sigma X;\ f_3 = \pi X - k_h P;\ \mu = \mu(S, X);\ \sigma = \sigma(S, X);\ \pi = \pi(S) \tag{43 – 48}$$

$$\frac{\partial f_1}{\partial X_1} = \mu + X\mu'_X;\ \frac{\partial f_1}{\partial X_2} = X\mu'_S;\ \frac{\partial f_1}{\partial X_3} = 0 \tag{49 – 51}$$

$$\frac{\partial f_2}{\partial X_1} = -(\sigma + X\sigma'_X);\ \frac{\partial f_2}{\partial X_2} = -X\sigma'_S;\ \frac{\partial f_2}{\partial X_3} = 0 \tag{52 – 54}$$

$$\frac{\partial f_3}{\partial X_1} = \pi;\ \frac{\partial f_3}{\partial X_2} = X\pi'_S;\ \frac{\partial f_3}{\partial X_3} = -k_h \tag{55 – 57}$$

Then

$$\frac{d\phi}{dt}=[\lambda_1\ \ \lambda_2\ \ \lambda_3]*\begin{bmatrix}-1 & 0 & 0\\ 0 & -1 & 0\\ 0 & 0 & -1\end{bmatrix}*\begin{bmatrix}f_1\\ f_2\\ f_3\end{bmatrix}-\begin{bmatrix}\mu+X\mu'_X & X\mu'_S & 0\\ -(\sigma+X\sigma'_X) & -X\sigma'_S & 0\\ \pi & X\pi'_S & -k_h\end{bmatrix}*\begin{bmatrix}g_1\\ g_2\\ g_3\end{bmatrix} \quad (58)$$

$$\frac{d\phi}{dt}=[\lambda_1\ \ \lambda_2\ \ \lambda_3]*\begin{bmatrix}-f_1\\ -f_2\\ -f_3\end{bmatrix}-\begin{bmatrix}g_1(\mu+X\mu'_X)+g_2X\mu'_S\\ -g_1(\sigma+X\sigma'_X)-g_2X\sigma'_S\\ g_1\pi+g_2X\pi'_S-g_3k_h\end{bmatrix} \quad (59)$$

$$\begin{aligned}\frac{d\phi}{dt}=&\lambda_1\Big(-f_1-g_1(\mu+X\mu'_X)-g_2X\mu'_S\Big)+\lambda_2\Big(-f_2+g_1(\sigma+X\sigma'_X)+g_2X\sigma'_S\Big)\\ &+\lambda_3(-f_3-g_1\pi-g_2X\pi'_S+g_3k_h)\end{aligned} \quad (60)$$

$$\begin{aligned}\frac{d\phi}{dt}=&\lambda_1\Big(-\mu X+X(\mu+X\mu'_X)+(S_f-S)X\mu'_S\Big)+\lambda_2\Big(\sigma X-X(\sigma+X\sigma'_X)\\ &+(S_f-S)X\sigma'_S\Big)+\lambda_3\Big(-(\pi X-k_hP)+\pi X-(S_f-S)X\pi'_S-k_hP\Big)\end{aligned} \quad (61)$$

$$\frac{d\phi}{dt}=\lambda_1\Big(\mu'_XX^2-(S_f-S)X\mu'_S\Big)+\lambda_2\Big(-\sigma'_XX^2+(S_f-S)X\sigma'_S\Big)+\lambda_3\Big(-(S_f-S)X\pi'_S\Big) \quad (62)$$

In the singular interval:

$$\frac{d\phi}{dt}=0 \quad (63)$$

$$H_0(t)=\lambda_1f_1+\lambda_2f_2+\lambda_3f_3=0 \quad (64)$$

$$\phi(t)=\lambda_1g_1+\lambda_2g_2+\lambda_3g_3=0 \quad (65)$$

From Eq. (64):

$$\lambda_1f_1+\lambda_2f_2+\lambda_3f_3=0\Rightarrow\lambda_1=-\frac{(\lambda_2f_2+\lambda_3f_3)}{f_1} \quad (66)$$

Substituting Eq. (66) into Eq. (65) results in the following equation:

$$\begin{aligned}&-\frac{(\lambda_2f_2+\lambda_3f)}{f_1}g_1+\lambda_2g_2+\lambda_3g_3=0\Rightarrow-\frac{\lambda_2f_2}{f_1}g_1-\frac{\lambda_3f_3}{f_1}g_1+\lambda_2g_2+\lambda_3g_3=0\Rightarrow\\ &\left(g_2-\frac{f_2}{f_1}g_1\right)\lambda_2=\left(\frac{f_3}{f_1}g_1-g_3\right)\lambda_3\Rightarrow\lambda_2=\frac{\left(\frac{f_3}{f_1}g_1-g_3\right)\lambda_3}{\left(g_2-\frac{f_2}{f_1}g_1\right)}\Rightarrow\lambda_2=\underbrace{\left(\frac{f_3g_1-f_1g_3}{f_1g_2-f_2g_1}\right)}_{\beta}\lambda_3\end{aligned} \quad (67)$$

By introducing the expression of λ_2 into the expression of λ_1, one obtains

$$\lambda_1=-\underbrace{\left[\frac{\left(\frac{f_3g_1-f_1g_3}{f_1g_2-f_2g_1}\right)f_2+f_3}{f_1}\right]}_{\alpha}\lambda_3 \quad (68)$$

Substituting $\lambda_1=\alpha\lambda_3$ and $\lambda_2=\beta\lambda_3$ into the expression of $d\phi/dt$ in the singular interval provides

$$\frac{d\phi}{dt}=\lambda_3\alpha\left(\mu'_X X^2-(S_f-S)X\mu'_S\right)+\lambda_3\beta\left(-\sigma'_X X^2+(S_f-S)X\sigma'_S\right)+\lambda_3\left(-(S_f-S)X\pi'_S\right)=0 \tag{69}$$

$$\frac{d\phi}{dt}=\lambda_3\underbrace{\left[\alpha\left(\mu'_X X^2-(S_f-S)X\mu'_S\right)+\beta\left(-\sigma'_X X^2+(S_f-S)X\sigma'_S\right)+\left(-(S_f-S)X\pi'_S\right)\right]}_{Q(\underline{X})}=0 \tag{70}$$

$$\frac{d\phi}{dt}=\lambda_3 Q(\underline{X})=0\Rightarrow Q(\underline{X})=0 \tag{71}$$

The equation $Q(\underline{X})=0$ is the expression of the singular arc, which is independent of the adjoint variables λ_1, λ_2, λ_3. In the expression of $Q(\underline{X})$, the indicated derivatives are given by

$$\mu=\frac{\mu_m S}{BX+S}\Rightarrow\mu'_X=-\frac{\mu_m BS}{(BX+S)^2};\ \mu'_S=\frac{(BX+S)\mu_m-\mu_m S}{(BX+S)^2} \tag{72 – 73}$$

$$\pi=\frac{\pi_m S}{k_m+S+\frac{S^2}{k_i}}\qquad\Rightarrow\pi'_S=\frac{\left(k_m+S+\frac{S^2}{k_i}\right)\pi_m-\pi_m S\left(1+\frac{2}{k_i}S\right)}{\left(k_m+S+\frac{S^2}{k_i}\right)^2} \tag{74}$$

$$\sigma=\frac{\mu}{Y_{X/S}}+\frac{\pi}{Y_{P/S}}+m\Rightarrow\sigma'_X=\frac{\mu'_X}{Y_{X/S}};\ \sigma'_S=\frac{\mu'_S}{Y_{X/S}}+\frac{\pi'_S}{Y_{P/S}} \tag{75 – 76}$$

For the determination of the singular dilution rate (u_{sin}), one starts from the first derivative of ϕ as follows:

$$\frac{d\phi}{dt}=\lambda_3 Q(\underline{X})\Rightarrow\frac{d^2\phi}{dt^2}=\lambda_3\frac{d}{dt}Q(\underline{X})+Q(\underline{X})\frac{d\lambda_3}{dt} \tag{77}$$

$$\frac{d}{dt}\underline{\lambda}=-\frac{\partial H}{\partial\underline{X}}\Rightarrow\begin{bmatrix}d\lambda_1/dt\\ d\lambda_2/dt\\ d\lambda_3/dt\end{bmatrix}=-\begin{bmatrix}\partial H/\partial X_1\\ \partial H/\partial X_2\\ \partial H/\partial X_3\end{bmatrix} \tag{78}$$

where

$$H=H_0(t)+\phi(t)u=\lambda_1 f_1+\lambda_2 f_2+\lambda_3 f_3+(\lambda_1 g_1+\lambda_2 g_2+\lambda_3 g_3)u \tag{79}$$

Thus:

$$\frac{d\lambda_3}{dt}=-\frac{\partial H}{\partial X_3}=-\lambda_3\frac{\partial f_3}{\partial X_3}-\lambda_3\frac{\partial g_3}{\partial X_3}u \tag{80}$$

Substituting the expression of $d\lambda_3/dt$ into the expression of the second derivative of φ gives

$$\frac{d^2\phi}{dt^2} = \lambda_3 \frac{d}{dt} Q(\underline{X}) + Q(\underline{X})\left(-\lambda_3 \frac{\partial f_3}{\partial X_3} - \lambda_3 \frac{\partial g_3}{\partial X_3} u\right) \tag{81}$$

When $u = u_{sin}$, the second derivative of ϕ is zero, i.e., $d^2\phi/dt^2 = 0$. Thus,

$$\begin{aligned} \lambda_3 \frac{d}{dt} Q(\underline{X}) + Q(\underline{X})\left(-\lambda_3 \frac{\partial f_3}{\partial X_3} - \lambda_3 \frac{\partial g_3}{\partial X_3} u\right) = 0 \Rightarrow \frac{d}{dt} Q(\underline{X}) - Q(\underline{X}) \frac{\partial f_3}{\partial X_3} \\ = Q(\underline{X}) \frac{\partial g_3}{\partial X_3} u_{\sin} \Rightarrow u_{\sin} = \frac{\left(\frac{d}{dt} Q(\underline{X}) - Q(\underline{X}) \frac{\partial f_3}{\partial X_3}\right)}{\left(Q(\underline{X}) \frac{\partial g_3}{\partial X_3}\right)} \end{aligned} \tag{82}$$

The expression of u_{sin} determines how the dilution rate must be manipulated (varied) during the singular interval, being a function only of the state variables. When developed, the final expression obtained for u_{sin} is quite complex and extensive, which was implemented in computational programming language to evaluate the value of this variable over the singular interval.

The condition for stopping the integration of mass-balance equations during the period following the singular interval, which is conducted in batch mode ($u = 0$), is determined from the fact that for free final time (t_f) problems, how is the case, $H(t_f) = 0$. Thus

$$\begin{aligned} H = \underline{\lambda}^T \left(\underline{f}(\underline{X}) + \underline{g}(\underline{X})\, u\right) \underset{u=0}{\Rightarrow} H = \underline{\lambda}^T \underline{f}(\underline{X}) \Rightarrow H = \lambda_1 f_1 + \lambda_2 f_2 + \lambda_3 f_3 \\ \Rightarrow H = \lambda_1 \frac{dX}{dt} + \lambda_2 \frac{dS}{dt} + \lambda_3 \frac{dP}{dt} \end{aligned} \tag{83}$$

As the final conditions of the adjoint variables are $\lambda_1 = 0$, $\lambda_2 = 0$, and $\lambda_3 = 1$, the stopping condition at $t = t_f$ is $dP/dt = 0$.

Thus, the problem-solving algorithm consisted of the following steps:

1. Integrate the mass-balance equations with $u = 0$ (batch operation) until the values of the state variables satisfy $Q(\underline{X}) = 0$, being the instant that this occurs, the first switching time t_1;

2. From instant t_1, make $u = u_{\text{sin}}$ until the reactor volume reaches its maximum value, which corresponds to the second switching time t_2;

3. Starting from time t_2, return the operation of the reactor with $u = 0$ until the stop condition is reached in t_f.

From the data reported by Costa [20] and summarized in **Table 1**, the mass-balance equations were numerically integrated to determinate t_1, i.e., the instant at which $Q(\underline{X}) = 0$. According to the data presented in **Table 2**, this time instant can be determined as $t_1 = 28.74$ h, since $Q(\underline{X})$ changes its signal around this time value.

Using the computational program developed for the calculation of u_{sin}, it was verified that this control variable was practically constant during the singular interval, being 0.00213 h^{-1} a value quite representative.

Kinetic parameters/operating variables	Values
μ_m (h^{-1}); B (g-S/g-X)	1.1×10^{-1}; 6×10^{-3}
π_m (g g^{-1} h^{-1}); k_m (g/L); k_i (g/L); k_h (h^{-1})	4.0×10^{-3}; 1.0×10^{-4}; 1.0×10^{-1}; 1.0×10^{-2}
$Y_{X/S}$ (g/g); $Y_{P/S}$ (g/g); m(g g^{-1} h^{-1})	0.47; 1.2; 2.9×10^{-2}
X_0 (g/L); S_0 (g/L); P_0 (g/L); S_f (g/L)	1.3; 69.0; 0.0; 500.0
V_0 (L); V_f (L)	8.121; 10.0

Table 1. Values of kinetic parameters and operating variables used in the case study of the optimization of penicillin production in fed-batch reactor (source: [20]).

t (h)	X (g/L)	S (g/L)	P (g/L)	$Q(\underline{X})$
28.742	30.09	4.16×10^{-3}	1.39×10^{-2}	125799.24
28.743	30.09	3.06×10^{-3}	1.40×10^{-2}	−86336.84

Table 2. Data of the numerical integration of the mass-balance equations used to determine the first switching time (t_1) during the penicillin production in fed-batch reactor.

With respect to the second switching time (t_2), this was determined from u_{sin} and the initial and final (maximum) volumes as follows:

$$\frac{dV}{dt} = F = uV \Rightarrow \frac{dV}{V} = udt \Rightarrow \ln\left(\frac{V_f}{V_0}\right) = u\Delta t \Rightarrow \ln\left(\frac{10}{8.121}\right) = 0.00213\Delta t \Rightarrow \Delta t = 97.71 \text{ h} \quad (84)$$

Thus, the singular interval duration is of 97.71 h and the second switching time (t_2) is $t_2 = t_1 + \Delta t = 28.74 + 97.71 = 126.45$ h. The mass-balance equations were then integrated until this time, and it was verified that, at the end of the integration, the stop condition (*dP/dt=0*) had been satisfied, dispensing with the complementary period of batch operation and reaching a final product concentration (P) of 6.35 g/L.

2.1.1. *Simulation of the penicillin production bioprocess in fed-batch reactor under optimized conditions*

The simulation of the bioprocess under optimized conditions was performed using a computer program in FORTRAN language. For the numerical integration of the ordinary differential equations corresponding to the mass balances, the variable-step fourth-order Runge-Kutta-Gill method was used [17]. The full profiles of the state variables during the bioprocess are shown in **Figures 1** and **2**.

Due to the decoupling between biomass growth and product synthesis, this type of fermentation behaves as a biphasic process. Therefore, characteristic profiles of penicillin fermentation were obtained for the state variables as shown in **Figure 1**, i.e., a first phase of accumulation of the cell is observed in which the substrate is almost entirely consumed for this purpose, without associated product formation (trophophase). After this growth phase, the fed substrate is practically used for penicillin production since there is no further catabolic repression of the antibiotic synthesis due to the low levels of substrate concentration established in the

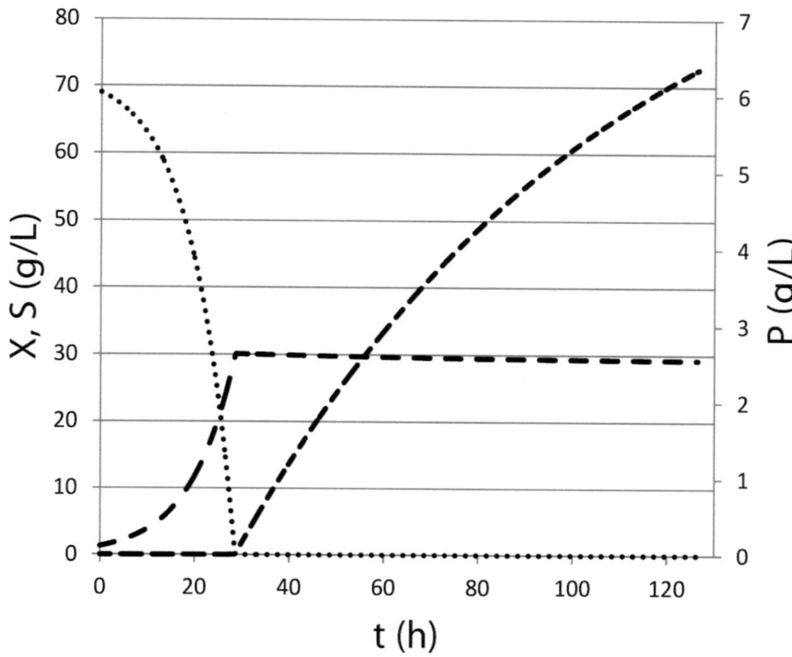

Figure 1. Temporal profiles of the *S* (dotted line), *X* (dashed line), and *P* (solid line) state variables during a fed-batch penicillin production bioprocess.

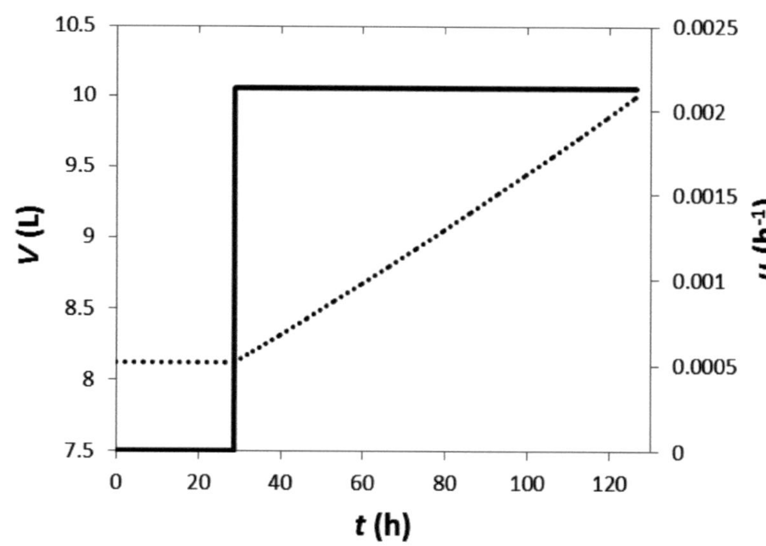

Figure 2. Temporal profile of the state variable *V* (dotted line) and of the control variable *u* (solid line) during a fed-batch penicillin production bioprocess.

reactor in this second phase (idiophase). In addition, the kinetic pattern observed is in agreement with that expected for a secondary metabolite, i.e., the production occurs mostly after cell growth.

Regarding the fermentation medium volume in the bioreactor, the behavior of this variable shown in **Figure 2** was already expected since, during the batch operation, this volume is constant because there is no addition or removal of fermentation medium to or from the bioreactor. In the fed-batch operation with continuous feed of unfermented medium to the bioreactor at a constant flow rate, the volume increases linearly over time, as shown in **Figure 2**. The temporal profile exhibited by the control variable u (dilution rate) in a step format is due to the change in the bioreactor operation, from batch ($u = 0$) to fed-batch ($u \neq 0$) mode.

2.2. Case study #2: determination of the optimal temperature profile in a batch bioreactor for penicillin production

In this case study, the fungal growth is described by the logistic law, a substrate-independent model for microorganism population dynamics. In addition, the production of penicillin is also modeled considering that the formation of antibiotic is not associated with cell growth, and that the product is degraded by hydrolysis according to a first-order kinetics. The mathematical model, comprising two ordinary differential equations corresponding to the mass balances of cell and product in a batch bioreactor, and containing four parameters, is represented by (more information about this model can be found in Ref. [21]):

$$\frac{dX}{dt} = r_X = \mu_m X\left(1 - \frac{X}{X_m}\right) \tag{85}$$

$$\frac{dP}{dt} = r_P - r_h = \beta X - k_h P \tag{86}$$

where t is the time, X is the cell concentration, P is the antibiotic concentration, r_X is the cell growth rate, r_P is the antibiotic production rate, r_h is the product's hydrolysis rate, and μ_m, X_m, β, and k_h are the parameters of the model, according to the following meanings: μ_m is the maximum specific growth rate, X_m is the maximum possible cell concentration to be achieved, β is the constant of product formation not associated with growth, and k_h is the rate constant for the antibiotic hydrolysis reaction.

For the application of the Pontryagin's maximum principle, the model variables were dimensionless and expressions describing the kinetic parameters (b_i) as a function of temperature (θ) were incorporated in order to extend the validity range of the model to non-isothermal conditions. These functions have shapes typical of those found in microbial or enzyme-catalyzed reactions (concave down parabolas). The hydrolysis of the antibiotic is neglected in this dimensionless version of the model, with this version given by the following equations [17]:

$$\frac{dy_1}{d\tau} = b_1 y_1 - \frac{b_1}{b_2} y_1^2, \; y_1(0) = 0.03 \tag{87}$$

$$\frac{dy_2}{d\tau} = b_3 y_1, \; y_2(0) = 0.0 \tag{88}$$

where

- y_1 = dimensionless concentration of cell (-); y_2 = dimensionless concentration of product (-); τ = dimensionless time, $0 \leq \tau \leq 1$ (-)

$$b_1 = w_1 \left[\frac{1.0 - w_2(\theta - w_3)^2}{1.0 - w_2(25 - w_3)^2} \right]; \; b_2 = w_4 \left[\frac{1.0 - w_2(\theta - w_3)^2}{1.0 - w_2(25 - w_3)^2} \right]; \; b_3 = w_5 \left[\frac{1.0 - w_2(\theta - w_6)^2}{1.0 - w_2(25 - w_6)^2} \right] \tag{89 – 91}$$

- $w_1 = 13.1; w_2 = 0.005; w_3 = 30°C$
- $w_4 = 0.94; w_5 = 1.71; w_6 = 20°C$

As in this case, $\underline{g}(\underline{X}) = 0$ and $u = 0$, due to the reactor being operated in a batch mode, the mass-balances equations are simplified to

$$\frac{d\underline{X}}{dt} = \underline{f}(\underline{X}) \tag{92}$$

where

$$\underline{X} = \begin{bmatrix} y_1 \\ y_2 \end{bmatrix}; \; \underline{f}(\underline{X}) = \begin{bmatrix} f_1 \\ f_2 \end{bmatrix} = \begin{bmatrix} b_1 y_1 - \frac{b_1}{b_2} y_1^2 \\ b_3 y_1 \end{bmatrix} \tag{93 – 94}$$

As previously established in the first case study, the Hamiltonian is given by

$$H = \underline{\lambda}^T \left(\underline{f}(\underline{X}) + \underbrace{\underline{g}(\underline{X})\, u}_{0} \right) \Rightarrow H = \underline{\lambda}^T \underline{f}(\underline{X}); \underline{\lambda}^T = [\lambda_1 \lambda_2] \Rightarrow$$
$$H = [\lambda_1 \lambda_2] \begin{bmatrix} f_1 \\ f_2 \end{bmatrix} = \lambda_1 f_1 + \lambda_2 f_2 = \lambda_1 \left(b_1 y_1 - \frac{b_1}{b_2} y_1^2 \right) + \lambda_2 (b_3 y_1) \tag{95}$$

The temporal variation rates of the adjoint variables λ_1 and λ_2 are formulated as

$$\frac{d\,\underline{\lambda}}{d\tau} = -\frac{\partial H}{\partial \underline{X}} \Rightarrow \frac{d}{d\tau} \begin{bmatrix} \lambda_1 \\ \lambda_2 \end{bmatrix} = \begin{bmatrix} -\lambda_1 b_1 + 2\frac{b_1}{b_2} y_1 - \lambda_2 b_3 \\ 0 \end{bmatrix} \tag{96}$$

From the previous equation, the following equations can be derived

$$\frac{d\lambda_1}{d\tau} = -b_1 \lambda_1 + 2\frac{b_1}{b_2} y_1 \lambda_1 - b_3 \lambda_2 \tag{97}$$

$$\frac{d\lambda_2}{d\tau} = 0 \tag{98}$$

The necessary condition for the optimization of the bioprocess is

$$\frac{\partial H}{\partial \theta}=0 \Rightarrow \frac{\partial H}{\partial \theta}=\lambda_1\left(y_1\left(\frac{\partial b_1}{\partial \theta}\right)-y_1^2\frac{\partial (b_1/b_2)}{\partial \theta}\right)+\lambda_2\left(y_1\frac{\partial b_3}{\partial \theta}\right)=0 \tag{99}$$

From the expressions $b_i = b_i(\theta)$, the following derivatives can be obtained

$$\frac{\partial b_1}{\partial \theta}=-\left[\frac{2w_1w_2(\theta-w_3)}{1.0-w_2(25-w_3)^2}\right];\ \frac{\partial (b_1/b_2)}{\partial \theta}=0;\ \frac{\partial b_3}{\partial \theta}=-\left[\frac{2w_5w_2(\theta-w_6)}{1.0-w_2(25-w_6)^2}\right] \tag{100 – 102}$$

By inserting the derivatives of the parameters with respect to the temperature into the expression of $\partial H/\partial \theta = 0$, the expression of the optimal temperature profile (θ_{opt}) is obtained as follows:

$$\theta_{opt}=\left[\frac{2\lambda_1 y_1 w_1 w_2 w_3}{1.0-w_2(25-w_3)^2}+\frac{2y_1 w_5 w_2 w_6}{1.0-w_2(25-w_6)^2}\right]\Big/\left[\frac{2\lambda_1 y_1 w_1 w_2}{1.0-w_2(25-w_3)^2}+\frac{2y_1 w_5 w_2}{1.0-w_2(25-w_6)^2}\right] \tag{103}$$

As previously demonstrated, when the objective is to maximize the antibiotic concentration at the end of the bioprocess, it is necessary that $\lambda_1(1) = 0$ and $\lambda_2(1) = 1$. Since $d\lambda_2/dt = 0$, the second condition requires that λ_2 be constant and equal to 1.0 in the entire time domain, i.e., $\lambda_2 = 1.0$ for $0 \le \tau \le 1$.

Several numerical methods have been developed to solve this two-point boundary-value problem arising from the application of the maximum principle of Pontryagin to a batch penicillin production bioprocess. Constantinides and Mostoufi [17] used the orthogonal collocation method to solve this problem, justifying that this method is more accurate than the finite difference method. The problem was solved here using a much simpler numerical method than that of the orthogonal collocation to integrate the differential equations, which is the variable-step fourth-order Runge-Kutta-Gill method [17]. Thus, the algorithm for solving the problem consisted of the following steps:

1. Assignment of an initial value for $\lambda_1(0)$;
2. Integration of the system of ODEs from $\tau = 0$ to $\tau = 1$ and verification if $\lambda_1(1) = 0$. If not, assign a new value to $\lambda_1(0)$ until the final condition is satisfied.

In order to make the computational algorithm autonomous for the determination of $\lambda_1(0)$, the Newton-Raphson method [17] was coupled to the numerical integration method, solving the following non-linear algebraic equation:

$$r\left(\lambda_1(0)\right)=[\lambda_1(1)]_{calculated}-\underbrace{[\lambda_1(1)]_{specified}}_{0}=0 \Rightarrow r\left(\lambda_1(0)\right)=[\lambda_1(1)]_{calculated}=0 \tag{104}$$

2.2.1. *Simulation of the penicillin production bioprocess in batch reactor under optimized non-isothermal conditions*

The proposed algorithm was implemented in FORTRAN programming language, and the profiles of the state variables (y_1, y_2, and θ) are presented in **Figures 3–5**. These profiles are in

http://dx.doi.org/10.5772/intechopen.69395

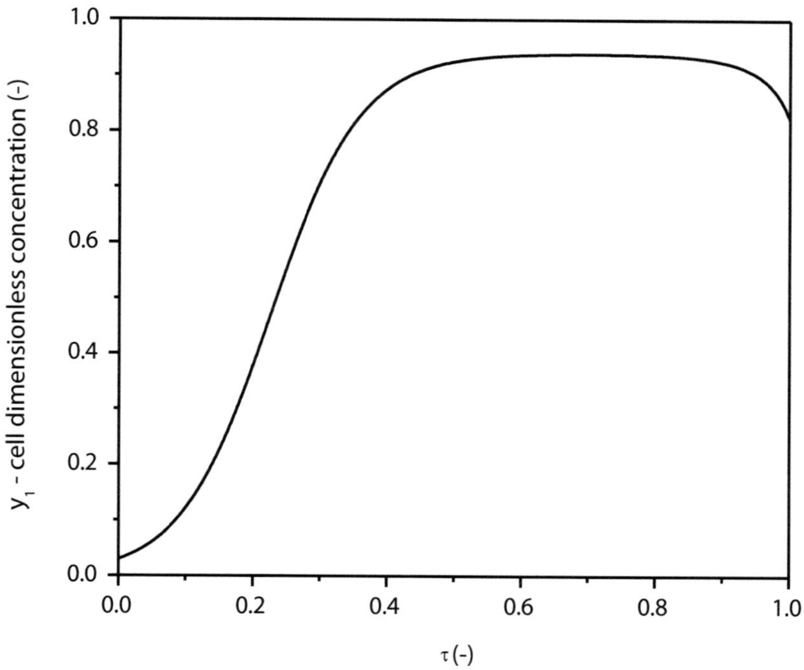

Figure 3. Cell dimensionless concentration profile during a non-isothermal penicillin fermentation.

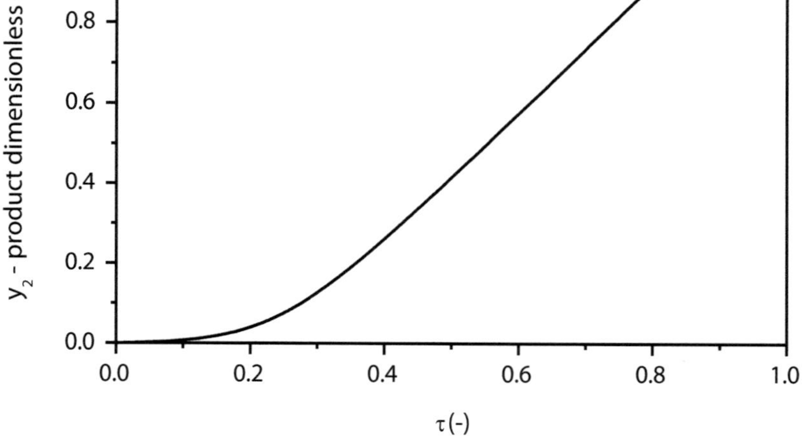

Figure 4. Product dimensionless concentration profile during a non-isothermal penicillin fermentation.

strict agreement with those reported by Constantinides and Mostoufi [17] when they used the orthogonal collocation method to solve this problem. The value determined for λ_1 at $\tau = 0$ was $\lambda_1(0) = 3.61210035$.

The cell concentration profile shown in **Figure 3** depicts the main phases involved in a typical microbial growth curve, i.e., the exponential, stationary, and decline phases. The decline phase is attributed to the negative effects of low temperatures on cell growth. In **Figure 4**, concerning the penicillin production dynamics, an initial short lag phase can be observed, followed by a transition phase in which penicillin production is initiated, until a final linear production phase is achieved.

According to the presented formulation (bioprocess model and Pontryagin's maximum principle), the optimum temperature profile varies between 20 and 30°C following the curve (a) shown in **Figure 5**. This profile suggests a variable operating temperature during the growth and penicillin production phases, contradicting the standard industrial operating procedure of maintaining a constant temperature throughout the bioprocess. Particularly during the penicillin production phase, the profile prescribes a decrease in the operating temperature so that a high concentration of antibiotic is reached at the end of the bioprocess. The temperature profile (a) shown in **Figure 5** may bring some practical difficulty to its programming/execution. In this context, an approximate profile derived from the exact profile, such as that represented by the curve (b) in **Figure 5**, may make the temperature programming strategy more feasible. This proposal is in agreement with that reported by Bailey and Ollis [22], i.e., the temperature schedule predicted by these calculations can be closely approximated in industrial practice with little added cost. The approximate profile was built from the following equations, which were based on the analysis of the temperature data generated by the exact profile:

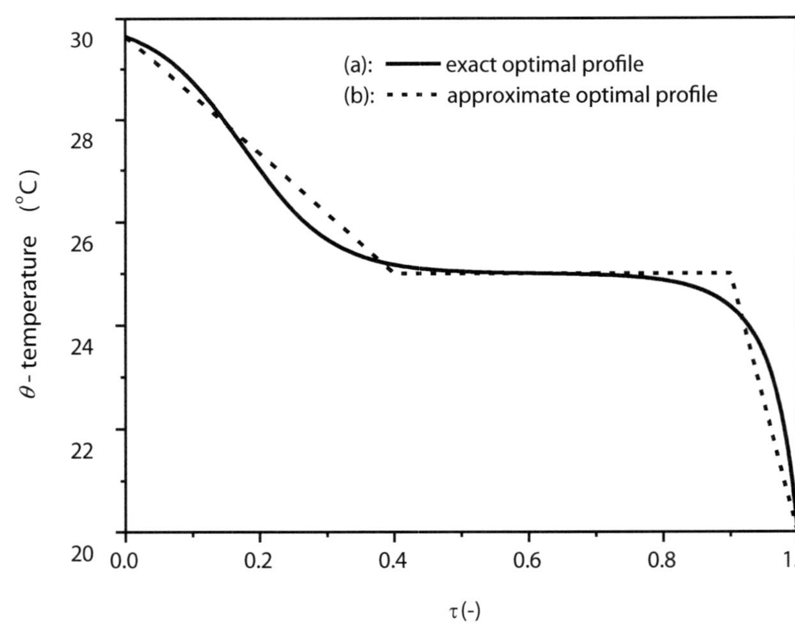

Figure 5. Exact and approximate optimal temperature profiles for a non-isothermal penicillin fermentation.

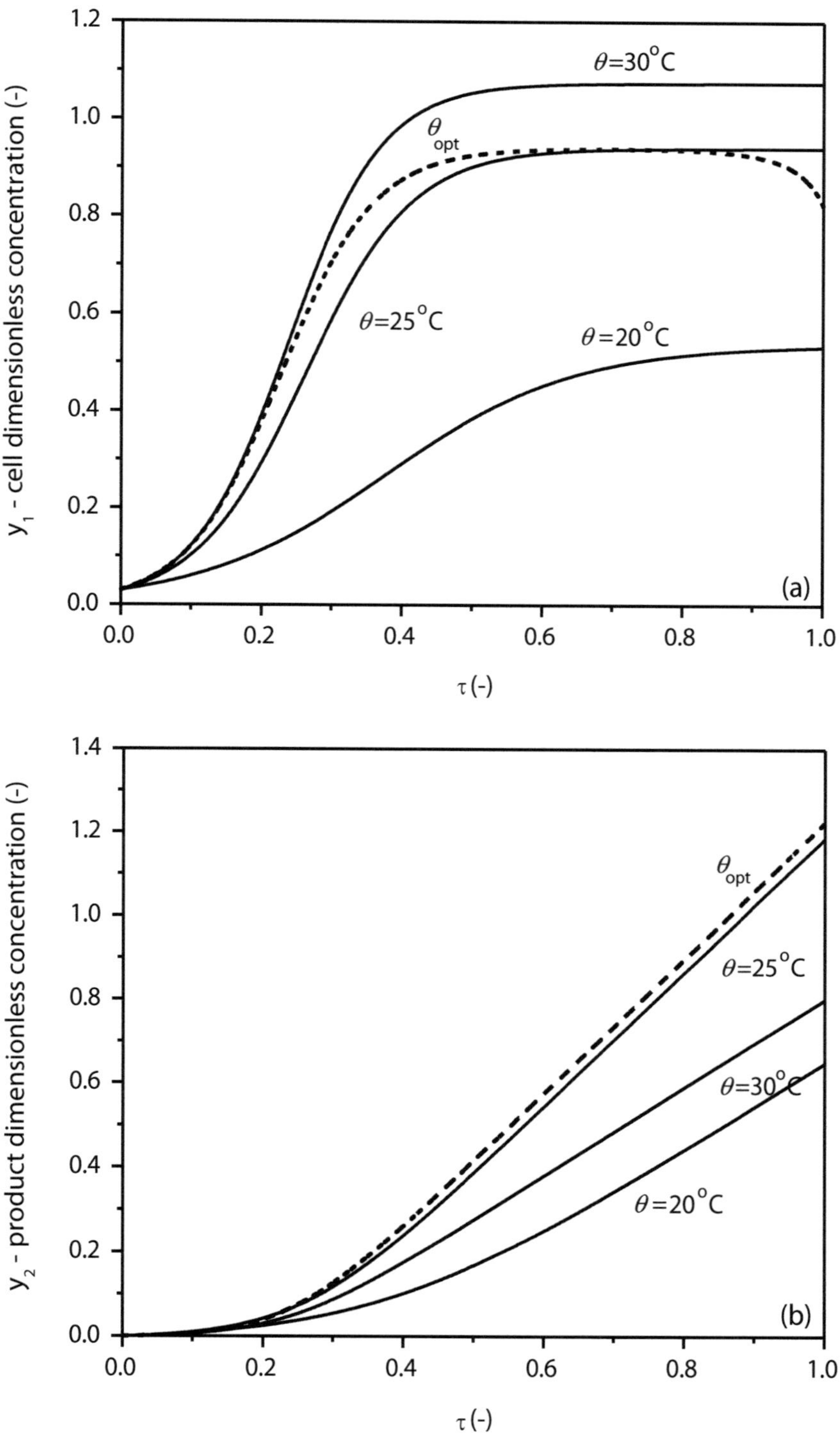

Figure 6. Dimensionless concentration profiles of cell and product during isothermal penicillin fermentations at different temperatures.

θ (°C)	y_1 (τ =1)	y_2 (τ =1)	$(y_2/y_1)_{\tau=1}$
θ_{opt}	0.82	1.22	1.49
20	0.53	0.65	1.23
25	0.94	1.18	1.25
30	1.07	0.80	0.75

Table 3. Final values of the dimensionless concentration of cells (y_1) and product (y_2) in isothermal and non-isothermal penicillin fermentations.

- $0.0 \le \tau \le 0.4$: $\theta(^\circ C) = \theta_{(\tau=0.0)} + \left(\frac{\theta_{(\tau=0.4)} - \theta_{(\tau=0.0)}}{0.4-0.0}\right)(\tau - 0.0) \Rightarrow \theta = 29.65 - 11.625\tau$
- $0.4 < \tau < 0.9$: $\theta(^\circ C) = 25.0$
- $0.9 \le \tau \le 1.0$: $\theta(^\circ C) = \theta_{(\tau=0.9)} + \left(\frac{\theta_{(\tau=1.0)} - \theta_{(\tau=0.9)}}{1.0-0.9}\right)(\tau - 0.9) \Rightarrow \theta = 25.0 - 50(\tau - 0.9)$

2.2.2. Simulation of the penicillin production bioprocess in batch reactor under non-optimized isothermal conditions

A pertinent simulation to be performed is one under isothermal conditions to verify whether this thermal operation mode is, in fact, less productive in penicillin than non-isothermal mode following an optimal temperature profile. For this purpose, simulations were performed for constant temperatures of 20, 25, and 30°C and the results were compared with those obtained with the optimized temperature profile (**Figure 6**). **Figure 6(a)** illustrates the well-known fact that high temperatures (30°C) favor the growth of the fungus, while low temperatures (20°C) favor the synthesis of the antibiotic since relative to the amount of penicillin produced, more and less biomass was accumulated at these respective temperature levels [22]. It is observed in **Figure 6(b)** that the isothermal operation at an intermediate temperature to those investigated (θ = 25°C) presents a performance very similar to the operation with optimized temperature profile, becoming a viable alternative if the variable temperature strategy cannot be implemented, although with lower productivity, according to the data presented in **Table 3**. Despite the fact that operation at a fixed temperature between 24 and 25°C predominates in current industrial practice, the benefits of temperature programming during batch antibiotic fermentations are clear.

3. Conclusions

In this chapter, the usefulness of the Pontryagin's maximum principle has been demonstrated for the optimization and operation of complex antibiotic production bioprocesses such as those conducted in batch and fed-batch reactors under isothermal/non-isothermal conditions. By applying this principle, it was possible to determine the optimal profile of temperature in batch reactors and substrate feed rate in fed-batch reactors that maximize the antibiotic concentration at the end of the bioprocess. Although having a rather complex mathematical formulation, the Pontryagin's maximum principle can be classified as a powerful and suitable tool for the optimization, control, and model-driven operation of bioprocesses aiming at maximum productivity of bioproducts. However, for the application of this principle, it is necessary to dispose a mathematical model, preferably phenomenological and representative of the bioprocess, in order to evaluate whether or not the solution found for a given problem is feasible. In the present study, two classical phenomenological models of penicillin production bioprocesses were used, together with the Pontryagin's maximum principle, aiming to determine the optimal operating conditions for the production of antibiotic, and the solutions found are considered feasible and can be implemented in real cases. However, a more complete mathematical model, incorporating the medium oxygenation state, could provide better bioprocess control, since the productivity in penicillin fermentations is highly dependent upon dissolved oxygen concentration, with its critical level being around 30% of saturation. In the models used here, the dissolved oxygen concentration was implicitly assumed to be non-limiting of the bioprocess, making this a rather restrictive hypothesis.

Acknowledgements

The author wishes to thank CNPq for their financial support (protocol number: 455487/2014-6).

Author details

Samuel Conceição de Oliveira

Address all correspondence to: samueloliveira@fcfar.unesp.br

Department of Bioprocesses and Biotechnology (DBB), School of Pharmaceutical Sciences (FCF), São Paulo State University (UNESP), Araraquara, SP, Brazil

References

[1] Sircar A, Sridhar P, Das PK. Optimization of solid state medium for the production of clavulanic acid by *Streptomyces clavuligerus*. Process Biochemistry. 1998;**33**(3):283-289

[2] Chen WC. Medium improvement for β-fructofuranosidase production by *Aspergillus japonicus*. Process Biochemistry. 1998;**33**(3):267-271

[3] Berkholz R, Guthke R. Model based sequential experimental design for bioprocess optimisation—An overview. In: Hofman M, Thonart P, editors. Engineering and Manufacturing for Biotechnology. Heidelberg, the Netherlands: Springer; 2002. pp. 129-141. DOI: 10.1007/0-306-46889-1

[4] Mandenius CF, Brundin A. Bioprocess optimization using design-of-experiments methodology. Biotechnology Progress. 2008;**24**:1191-1203

[5] Núñez EGF, Véliz RV, Costa BLV, Rezende AG, Tonso A. Using statistical tools for improving bioprocesses. Asian Journal of Biotechnology. 2013;**5**:1-20

[6] Kennedy M, Krouse D. Strategies for improving fermentation medium performance: A review. Journal of Industrial Microbiology & Biotechnology. 1999;**23**:456-475

[7] Lim HC, Lee KS. Process control and optimization. In: Pons MN, editor. Bioprocess Monitoring and Control. Munique: Hanser Publishers; 1992. pp. 159-222

[8] Chu WBZ, Constantinides A. Modeling, optimization, and computer control of the cephalosporin C fermentation process. Biotechnology and Bioengineering. 1988;**32**(3):277-288

[9] Lee J, Lee SY, Park S, Middelberg APJ. Control of fed-batch fermentations. Biotechnology Advances. 1999;**17**:29-48

[10] Chaudhuri B, Modak JM. Optimization of fed-batch bioreactor using neural network model. Bioprocess Engineering. 1998;**1**:71-79

[11] Zhang H, Zhang Z, Lan LH. Evolutionary optimization of a fed-batch penicillin fermentation process. In: 2010 International Symposium on Computer, Communication, Control and Automation; 5-7 May 2010; Taina. IEEE; 2010. pp. 403-406

[12] Rani KY, Rao VSR. Control of fermenters—A review. Bioprocess Engineering. 1999;**21**:77-88

[13] Ashoori A, Moshiri B, Khaki-Sedigh A, Bakhtiari MR. Optimal control of a nonlinear fed-batch fermentation process using model predictive approach. Journal of Process Control. 2009;**19**:1162-1173

[14] Rocha M, Mendes R, Rocha O, Rocha I, Ferreira EC. Optimization of fed-batch fermentation processes with bio-inspired algorithms. Expert Systems with Applications. 2014;**41**:2186-2195

[15] Roeva O, Tzonkov S. A genetic algorithm for feeding trajectory optimisation of fed-batch fermentation processes. Bioautomation. 2009;**12**:1-12

[16] Guthke R, Knorre WA. Optimal substrate profile for antibiotic fermentations. Biotechnology and Bioengineering. 1981;**23**:2771-2777

[17] Constantinides A, Mostoufi N. Numerical Methods for Chemical Engineers with MATLAB Applications. Upper Saddle River: Prentice Hall PTR; 1999. p. 560

[18] Van Impe JF, Nicolai BM, Vanrolleghem PA, Spriet JA, De Moor B, Vandewalle J. Optimal control of the penicillin G fed-batch fermentation: An analysis of the model of Heijnen *et al.* Optimal Control Applications & Methods. 1994;**15**:13-34

[19] Skolpap W, Scharer JM, Douglas PL, Moo-Young M. Optimal feed rate profiles for fed-batch culture in penicillin production. Songklanakarin Journal of Science and Technology. 2005;**27**(5):1057-1064

[20] Costa AC. Singular control in bioreactors [thesis]. Rio de Janeiro: Federal University of Rio de Janeiro (UFRJ); 1996

[21] Oliveira SC. Mathematical modeling of batch antibiotic production: Comparative analysis between phenomenological and empirical approaches. Journal of Bioprocessing & Biotechniques. 2017;7(1):298. DOI: 10.4172/2155-9821.1000298

[22] Bailley JE, Ollis DF. Biochemical Engineering Fundamentals. 2nd ed. New York, NY: McGraw-Hill; 1986. p. 984

An Overview of Response Surface Methodology Approach to Optimization of Hydrogen and Syngas Production by Catalytic Reforming of Greenhouse Gases (CH_4 and CO_2)

Bamidele V. Ayodele and Sureena Abdullah

Additional information is available at the end of the chapter

http://dx.doi.org/10.5772/intechopen.73001

Abstract

Catalytic reforming of Methane (CH_4) and carbon dioxide (CO_2) is one of the techniques used for the production of hydrogen and syngas. This technique has dual advantages of mitigation of greenhouse gases and production of hydrogen and syngas which are often used as intermediates for the synthesis of valuable chemical products and oxygenates. This study presented an overview of the application of response surface methodology (RSM) in the optimization of hydrogen and syngas production from catalytic reforming of CH_4 and CO_2. The different catalytic system that has been employed together with the nature of experimental design, input parameters, responses, the optimum conditions and the maximum values of their responses were examined. The future research direction in the application of RSM to optimization of hydrogen and syngas production by catalytic reforming of CH_4 and CO_2 was recommended.

Keywords: greenhouse gases, response surface methodology, catalytic reforming, hydrogen, syngas

1. Introduction

The frequent environmental pollution often encountered from the consumption of energy derived from fossil fuel has aroused the quest for production of alternative and cleaner source of energy [1, 2]. One of such alternative means of energy production is catalytic methane dry reforming whereby the two principal greenhouse gases, carbon dioxide (CO_2) and methane (CH_4) are utilized for the production of hydrogen and syngas using active catalysts [3–5]. Compare to other forms of reforming processes which utilized steam and oxygen; the dry methane reforming has a dual advantage of mitigating greenhouse effect by utilizing the two principal components of greenhouse gases CH_4 and CO_2 as feedstocks. Besides, hydrogen and syngas are produced which can either be used directly as fuel or as a chemical intermediate for the synthesis of value-added chemicals and synthetic fuel [6–8]. Thermodynamically, the dry methane reforming reaction requires temperature >500°C to be feasible [9, 10]. Hence, at a temperature >500°C, coke formation and deposition on the catalyst surface is often induced mainly by methane cracking and Boudouard reactions [11, 12]. The deposited coke usually lead to deactivation of the catalyst thereby reducing the activities and stability [13–15]. To overcome these major challenges, several supported catalysts such as Ni, Co, Pt, Pd, Ru, Rh, Li and so on have been employed to catalyze the production of hydrogen and syngas via dry methane reforming [14, 16, 17]. The findings from these studies revealed that the individual catalysts displayed different degrees of catalytic activities and stabilities during the dry methane reforming reaction. Hence, obtaining a consensus on the optimum conditions that can maximize the hydrogen and the syngas yield has been a bone of contention till date. In view of this, several authors have employed response surface methodology approach to investigate the optimum conditions required for obtaining maximum hydrogen and syngas yield from catalytic methane dry reforming. This study, therefore, presents an overview of the different response surface methodology (RSM) approaches that have been used to optimize hydrogen and syngas production by methane dry reforming using different catalysts.

2. Metal-based dry reforming catalysts

An extensive review by Pakhare and Spivey [14], Budiman et al. [16] and Kathiraser et al. [18] revealed that supported metal catalysts such as Pt, Rh, Ru, Co, and Ni had been widely investigated for hydrogen and syngas production by dry methane reforming. According to Pakhare and Spivey [14] noble metals such as Pt, Rh and Ru displayed high catalytic activities and stabilities towards dry reforming of methane even with minimal metal loading. However, the activities of these noble metals were observed to vary with the nature of support. Supports such as SiO_2, La_2O_3, ZrO_2, TiO_2, CeO_2, Al_2O_3, and MgO have been investigated for these noble metals [14, 19–21]. The significant physicochemical properties that influence the activities of the noble metals-based catalyst are their high metal dispersion and the metal surface area. Nevertheless, noble metals are pricey and not readily available. Hence, their usage might not be economical in the eventuality of a scale-up process.

In view of this, other supported metal catalysts mostly Co, and Ni have been given full attention as catalytic candidates for hydrogen and syngas production by dry methane reforming [16, 18]. Although several studies have shown that Co and Ni catalysts have inferior catalytic activities and stability compared to the noble metals, these metals are inexpensive and readily available. Hence, Co and Ni-based catalysts have been tipped as potential candidates for scale-up of catalytic dry methane reforming. Moreover, the catalytic properties of the Co and Ni catalysts can be improved to be competitive with that of the noble metals by either using suitable supports or promoters [14].

3. RSM approach to process optimization

Chemical process optimization is an important activity performed on a system or a process to obtain optimum conditions that can give the maximum benefit from such process. This can be achieved statistically by using one-variable-at-a-time for optimizing a response [22]. This method of optimization entails changing the parameters of a variable while keeping the level of the other variables constant [23]. One major drawback of this type of optimization is that the interaction effects of the variables are not usually considered during the optimization process [22]. Hence, the one-variable-at-time technique does not capture the broad effects of the parameters on the responses [24]. Besides, the technique requires a massive number of experimental runs which invariably implies an increase in time of experiment as well as high cost of reagents and materials [22].

The challenges using the one-variable-at-time form of optimization of the chemical process can be overcome using response surface methodology (RSM). RSM is a more robust optimization technique that presents the statistical design of experiment (DoE) which can be employed in achieving process with ultimate performance. RSM as a technique for chemical process optimization is made up of a set of mathematics and statistical tools which work based on empirical model fittings to the obtained experimental data from DoE [25]. The empirical model fittings help in developing a suitable functional relationship between a set of input variables and the targeted response [25]. The different stages involved in the use of RSM for chemical process optimization are depicted in **Figure 1** [22]. These stages include screening of the variables that have been identified for optimization, the choice of the experimental design, the codification of the level of the variables, mathematic-statistic treatment of data, evaluation of fitted model and determination of the optimum conditions.

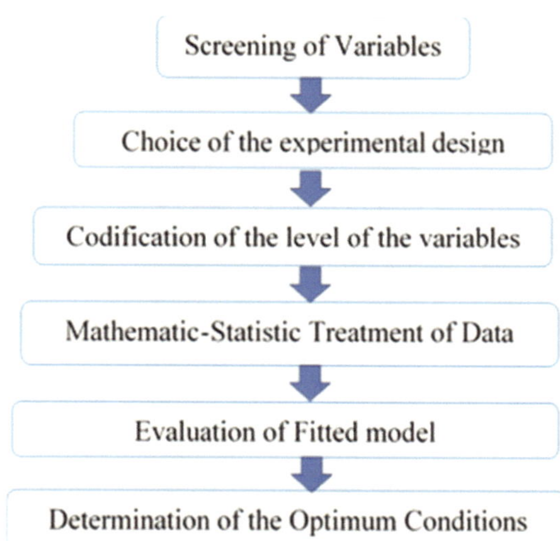

Figure 1. Stages involved in RSM applications [22].

Several parameters are often needed to investigate chemical processes extensively. One of the most significant challenges is to investigate all the effects of these parameters on the process. An attempt to do this might be rigorous, time-consuming and expensive [26]. Hence, it is expedient to determine the parameters that significantly affect the responses from the chemical process. To achieve this, experiment to identify parameters with most significant effects is usually performed at the preliminary stage using factorial designs [26]. Screening of variable before the main DoE has the advantages of preventing the mistake of chosen wrong levels that might negatively influence overall success of the process optimization. Having ascertained the appropriate parameters to be used for the main experiment, the next stage is to choose the right experimental design.

Experimental design can be selected based on the intention of using a simple model which can be employed as a linear function depicted in Eq. (1)

$$y = \beta_o \sum_{i=1}^{k} \beta_i x_i + \varepsilon \tag{1}$$

where k, β_o, β_i, x_i and ε depicts number of variables, the constant term, coefficients of the liner parameters, the input variables and the residual associated with the experiments, respectively. The response obtained from the linear model cannot be used to determine curvature. Hence the need for a second-order model represented in Eq. (2).

$$y = \beta_o + \sum_{i=1}^{k} \beta_i x_i + \sum_{1 \leq i \leq j}^{k} \beta_{ij} x_j + \ \varepsilon \tag{2}$$

where the coefficient of the interaction parameter is denoted as β_{ij}. The critical point which could either be maximum, minimum or saddle can be determined by including quadratic terms to the polynomial terms in Eq. (2) as shown in Eq. (3).

$$y = \beta_o + \sum_{i=1}^{k} \beta_i x_i + \sum_{1z \leq x_j\ i \leq j}^{k} \beta_{ij} + \ \sum_{1=i}^{k} \beta_{ii} x_i^2 + \ \varepsilon \tag{3}$$

where the coefficient of the quadratic parameter is denoted as β_{ii}.

Eq. (3) can be identified as a form two-modeling, symmetrical response surface designs which could be three-level factorial design, Box-Behnken design (BBD), central composite design (CCD), and Doehlert design. The main difference between these two modeling, symmetrical response surface designs is the selection of their experimental points, number of levels for variables, and the number of runs and block [22].

After choosing the appropriate experimental design, the next stage is the codification of the level of the variables which entail the transformation of the process real values to a coordinate within dimensionless value scale proportional to the localization in the experimental space [22]. One significant advantage of codification is that it allows the variables of different orders of magnitude to be determined without substantial influence on the lesser values.

The data obtained for each of the experimental point based on the selected experimental design can be subjected to mathematical-statistical treatment. The mathematical-statistical treatment entails the fitting of an appropriate mathematical equation that can best describe the behavior of the responses. The method of least squares which is a statistical approach can also be employed to fit a mathematical model to a given set of experimental data [27]. The mathematic model obtained from the treatment of the data can subsequently be evaluated to determine if it appropriately explains the experimental sphere investigated. This can be achieved by employing the analysis of variance (ANOVA). The use of ANOVA enables the comparison between the variations that arise from the treatment of the experimental data and the variation as a result of the random errors that accompanied the measurement of the obtained responses [28]. Moreover, the ANOVA helps to determine the significance and the mathematical model adequacy [28]. Besides ANOVA, other tools such as normality test, regression analysis and lack of fit test can be employed to examine the model adequacy of the RSM optimization. The normality of the experimental data can be performed using the normal plot of the internally studentized residuals. The data points in a normal plot is linear when the studentized residuals are normally distributed. Otherwise, when the data points in a normal plot is non-linear, it implies that the studentized residuals are not normally distributed. Hence, there is a need for the correction of the responses. Regression analysis is often performed on the fittings of the model denoted by the equation to the experimental data. The regression analysis of the model helps to determine to what extent the fitted model accounted for the variations in the experimental data. In order to further test the adequacy of the specified model, the lack of fits test can be employed. A significant lack of fit implies that the specific model is not suitable to explain the experimental data. Hence, a different form of model would adequately fit the data if investigated. The last stage of the application of the RSM is the determination of the optimum conditions that can maximize the response values. Numerical optimization using the RSM techniques can be employed to obtain the desired value for each of the input variables as a function of the target response. This is dependent on the input optimization strategies such as the range, maximum, minimum or target set to obtain the maximum achievable desired responses of the chemical process [29].

4. Optimization of hydrogen and syngas production using RSM

The details of the optimization studies on the catalytic reforming of CH_4 and CO_2 to hydrogen and syngas is depicted in **Table 1**. It can be seen that there is a dearth of literature on the optimization of hydrogen and syngas by the catalytic reforming of CH_4 and CO_2 despite the volume of literature available on their catalytic activities, stabilities and kinetic studies as reported by Budiman et al. [16], Pakhare and Spivey [14], and Kathiraser et al. [18]. The studies show that supported Ni-based catalysts such as Ni/γ-Al_2O_3, and Ni/SiO_2 has been employed for investigating the optimum conditions of hydrogen and syngas production from reforming of CH_4 and CO_2 using CCD. CCD as an RSM technique entails the use of a full factorial or fractional factorial design, a star design which consist of experimental points at a distance α from its center and the central point [30, 34]. For the Ni/γ-Al_2O_3 catalyst, the effect of factors such as discharge power, total flow rate CO_2/CH_4 molar ratio and Ni-loading on responses such as CO_2 conversion, CH_4 conversion, and CO yield, H_2 yield and the fuel production efficiency were investigated [30]. The ANOVA results show that all the factors investigated

RSM approach	Catalyst	Factors	Low-level	High-level	Responses	Optimum conditions		References
						Factors	Responses	
CCD	Ni/γ-Al_2O_3	Discharge power	30	50	CO_2 conversion	Discharge power = 60 W	CO_2 conversion = 31.7%	[30]
		Total flow rate	50	100	CH_4 conversion	Total flow rate = 56.1 mL/min	CH_4 conversion = 48.1%	
		CO_2/CH_4 molar ratio	0.75	1.25	CO yield (%)	CO_2/CH_4 molar ratio = 1.03	CO yield (%) = 21.7%	
		Ni loading	7.5	12.5	H_2 yield (%)	Ni loading = 9.5%	H_2 yield (%) = 17.9%	
					Fuel production efficiency (%)		Fuel production efficiency (%) 7.9%	
CCD	Li/MgO	Reaction temperature	650	800	% CH_4 conversion	Temperature = 725°C	% CH_4 conversion = 41.4	[31]
		GHSV (F/W)	5140	12,000	% C_2 selectivity	GHSV (F/W) = 8570 $cm^3\ g^{-1}\ h^{-1}$	% C_2 selectivity = 77.2	
		Li loaded	0.05	0.15	% C_2 yield	Li loaded = 0.1	% C_2 yield = 32	
CCD	15wt%Rh/MgO	Reaction temperature	650	850	% CH_4 conversion	Temperature = 918°C	% CH_4 conversion = 93.91720	[32]
		O_2/CH_4 ratio	0.1	0.2	% H_2 selectivity	O_2/CH_4 ratio = 0.15	% H_2 selectivity = 34.55	
		Catalyst weight	100	300	H_2/CO product ratio	Catalyst weight = 200	H_2/CO product ratio = 1.42	
CCD	Ni–Co/MSN	Temperature	700	800	% CH_4 conversion	Temperature = 783°C	% CH_4 conversion = 97	[33]
		CO_2/CH_4 ratio	1	5		CO_2/CH_4 ratio = 3		
		GHSV	10,000	60,000		GHSV = 38,726 mL $g^{-1}\ h^{-1}$.		
CCD	Ni/SiO_2	Reaction temperature	600	800	CH_4 conversion	Reaction temperature = 800°C	CH_4 conversion = 79.6	[34]
		CH_4/CO_2 molar ratio	0.25	4	CO_2 conversion	CH_4/CO_2 molar ratio = 2.125	CO_2 conversion = 84.2	
					The H_2/CO ratio		H_2/CO ratio = 0.4	
					Carbon content		Carbon content = 51.1	

http://dx.doi.org/10.5772/intechopen.73001

RSM approach	Catalyst	Factors	Low-level	High-level	Responses	Optimum conditions		References
						Factors	Responses	
BBD	Fe/Mg/ Al_2O_3	Reaction temperature	550	750	CH_4 conversion	Reaction temperature = 650°C	CH_4 conversion = 100%	[35]
		Mg loading	0	10	% H_2 yield	Mg loading = 5	% H_2 yield = 83.7%	
		Synthesis method	Co-impregnation	Sequential impregnation	CO/CO_2 molar ratio	Synthesis method = Sequential impregnation	CO/CO_2 molar ratio = 16.5%	
CCD	Ni-Co/MgO-ZrO_2	CO_2/CH_4 ratio	1	5	CH_4 conversion	CO_2/CH_4 ratio = 3	CH_4 conversion = 88 mol%	[30]
		GHSV	8400	200,000	H_2 yield (%)	GHSV = 145,190 mL g^{-1} h^{-1}	H_2 yield (%) = 86 mol%	
		Oxygen concentration in the feed	3	8		Oxygen concentration in the feed = 7 mol%		
		Reaction temperature	700	800		Reaction temperature = 749°C		
CCD	ZnO	Weight of catalyst	5	8		Weight of catalyst = 8 g	CH_4 conversion = 11.88%	[36]
		Total pressure	30	90		Total pressure = 30 Psi	CO_2 conversion = 15.69%	
		CO_2:CH_4:He (% CO_2)	10	80		CO_2:CH_4:He (% CO2) = 10%		
		UV light power	80	250		UV light power = 250 W		
BBD	Co/Sm_2O_3	CH_4 partial pressure	10	50	H_2 yield (%)	CH_4 partial pressure = 47.9 kPa	H_2 yield (%) = 79.4	[37]
		CO_2 partial pressure	10	50	CO yield (%)	CO_2 partial pressure = 48.9 kPa	CO yield (%) = 79.0	
		Reaction temperature	650	750		Reaction temperature = 735°C		

RSM approach	Catalyst	Factors	Low-level	High-level	Responses	Optimum conditions		References
						Factors	**Responses**	
BBD	Co/CeO_2	CH_4 partial pressure	10	50	CH_4 conversion (%)	CH_4 partial pressure = 46.85		
		Reaction temperature	650	750	CO_2 conversion (%)	Reaction temperature = 727°C	CH_4 conversion (%) = 74.85	[38]
		CO_2/CH_4 ratio	0.4	1	Syngas ratio	CO_2/CH_4 ratio = 0.6	CO_2 conversion (%) = 76.49	
							Syngas ratio = 0.97	

Table 1. Summary of literature on optimization of hydrogen and syngas production from catalytic reforming of CH_4 and CO_2 using RSM.

have significant effects on the responses since their p-values were <0.05. Optimum conditions of 60 W, 56.1 mL/min, 1.03, and 9.5% were obtained for the power discharged, total flow rate, CO_2/CH_4 molar ratio, and Ni-loading, respectively. These optimum conditions resulted in maximum values of 31.7%, 48.1%, 21.7%, 17.9%, and 7.9% of CO_2 conversion, CH_4 conversion, CO yield, H_2 yield, and fuel production efficiency, respectively. Similarly, for the Ni/SiO_2 catalyst, the effect of input variables such as reaction temperature and CH_4/CO_2 molar ratio on the CH_4 conversion, CO_2 conversion, H_2/CO ratio and carbon content were investigated using CCD [34]. Based on the ANOVA results, the factors investigated were observed to have a significant effect on the CH_4 conversion, CO_2 conversion, H_2/CO ratio and carbon content ($p < 0.05$). The optimum process conditions of 800°C and 2.125 were obtained for reaction temperature and CH_4/CO_2 molar ratio, respectively yielding maximum values of 79.6%, 84.2%, 0.4 and 51.1% for CH_4 conversion, CO_2 conversion, H_2/CO ratio and carbon content, respectively.

In addition to supported Ni catalysts, other Ni-containing bimetallic catalysts such as Ni–Co/MSN, and Ni-Co/MgO-ZrO_2 catalyst have been employed to study the effect of different process variables on their respective responses using CCD [30, 33]. The variables investigated include reaction temperature, CO_2/CH_4 ratio, GHSV, and O_2 concentration in the feed while the responses include CH_4 conversion and H_2 yield. The ANOVA results of the fittings of the experimental data obtained using both the Ni-Co/MSN, and Ni-Co/MgO-ZrO_2 catalysts showed that the input variables had significant influence on the responses. The reforming of CH_4 and CO_2 over the Ni-Co/MSN gave optimum conditions of 783°C, 3, and 38,726 mL g^{-1} h^{-1} for the reaction temperature, CO_2/CH_4 ratio, GHSV, respectively yielding maximum value 97% for the CH_4 conversion. Similarly, the reforming of CH_4 and CO_2 over the Ni-Co/MgO-ZrO_2 catalyst gave optimum conditions of 749°C, 3, 145,190 mL g^{-1} h^{-1}, and 7 mol% for the reaction temperature, CO_2/CH_4 ratio, GHSV O_2 concentration in the feed, respectively. Consequently, maximum values of 88% and 86% for CH_4 conversion and H_2 yield, respectively were obtained at the optimum conditions.

The optimization of hydrogen and syngas production from catalytic reforming of CH_4 and CO_2 over Sm_2O_3 and CeO_2 supported Co catalysts have been investigated using BBD [37, 38]. The BBD is more efficient and less costly compared to the three-factor design. This is because the BBD allows an effective estimate of the first and second-order coefficients of the mathematical model [22]. The effects of process factors such as reaction temperature, CH_4 partial pressure, CO_2 partial pressure, and CO_2/CH_4 ratio on H_2 yield, CO yield, CH_4 conversion, and CO_2 conversion were investigated using both Co/Sm_2O_3 and Co/CeO_2 catalysts. The p-value (<0.05) obtained from the ANOVA results of the study revealed that all the factors investigated significantly influence the responses. The reforming of CH_4 and CO_2 over the Co/Sm_2O_3 catalyst led to optimum conditions of 727°C, 47.9 kPa, 48.9 kPa for the reaction temperature, CH_4 partial pressure, and CO_2 partial pressure, respectively leading to the maximum values of 79.4% and 79% for H_2 yield, and CO yield respectively. Likewise, the reforming of CH_4 and CO_2 over the Co/CeO_2 catalyst resulted in optimum conditions of 727°C for the reaction temperature, 46.85 kPa for the CH_4 partial pressure, and 0.6 for the CO_2/CH_4 ratio. These optimum conditions resulted in maximum values of 74.85% for CH_4 conversion, 76.49% for CO_2 conversion and 0.97 for the syngas ratio. Besides Ni and Co-based catalysts, other catalysts such as Li/MgO, 15wt%Rh/MgO, Fe/Mg/Al_2O_3, and ZnO catalysts have been investigated for optimization of hydrogen and syngas production from reforming of CH_4 and CO_2 [31, 32, 35, 36]. The ANOVA results obtained from these studies indicate that the all the factors investigated had significant effects on their responses.

5. Implications for further research

The overview of RSM approach to the optimization of hydrogen and syngas from reforming of CH_4 and CO_2 over different catalysts performed in this study has revealed that every reforming catalyst displayed a unique set of optimum conditions. This trend might be due to the temperature dependent nature of the reforming reaction and the unique physicochemical properties of each of the catalysts investigated. As a result, there is no consensus on unified optimum conditions for hydrogen and syngas production by catalytic reforming of CH_4 and CO_2. Moreover, the study also shows that only CCD and BBD has been employed for the optimization of the hydrogen and syngas production over the catalysts investigated. Hence, other forms of experimental design such as Doehlert design and Taguchi can further be explored for the optimization study and then compare with existing one in literature. Furthermore, none of the work reported in this study perform an initial screening of the variable. A school of thought has argued that screening of variable is only essential when using Plackett-Burman design. However, it is worthwhile investigating the effect of initial screening of the different variables that can potentially influence hydrogen and syngas production from reforming of CH_4 and CO_2. Moreover, to ensure reliability and accuracy, high fractional factorial design can be employed during the pre-screening stage. An efficient pre-screening of all possible factors that influence the production of hydrogen and syngas by dry methane reforming will enable the most significant factors to be obtained for subsequent optimization. A consensus can be arrived at using these significant factors for further optimization of the hydrogen and syngas production with different catalytic system.

Although, response surface design is usually carried out using continuous factors (factors whose values are fixed to investigate their relationship to a response), it will be worthy of investigation is research efforts can be geared towards response surface design using categorical factors since such factors have discrete settings with no specific order. This could help eliminating the discrepancy in the variations in the optimum conditions obtained using different catalytic system in the optimization reforming reactions.

6. Conclusion

There is a growing interest in the application of RSM for the optimization of chemical processes due to its numerous advantages over the traditional one-variable-a-time optimization techniques. Such advantages include the tendency to obtain a large aggregate of information from a small set of experimental runs and propensity to determine the effect of interaction between the variables on the process responses. In this overview, RSM as an optimization technique has been applied for optimization of catalytic reforming of CH_4 and CO_2 over few catalytic systems mainly supported Ni and Co catalysts as well as their bi-metallic components. This study also affirms that the choice of the experimental design employed for the optimization process was limited to CCD and BBD. Despite that all the literature considered in this study reported the significant interactive effect of the input variables on their respective responses, there was, however, no consensus of the optimum conditions of all the catalysts investigated. Each of the catalysts was observed to yield a unique set of optimum conditions primarily due to the differences in their physicochemical properties. This study has revealed that there is dearth literature in the application of RSM to optimization of hydrogen and syngas production from the catalytic reforming of CH_4 and CO_2. Hence, it is therefore recommended that research efforts should be concentrated on investigating other RSM techniques besides BBD and CCD as well as considering using more catalytic system for the future optimization study.

Acknowledgements

Bamidele Ayodele Victor is a grateful recipient of Universiti Malaysia Pahang Post-Doctoral Fellowship. This work is supported by the Universiti Malaysia Pahang internal grant (RDU15097).

Author details

Bamidele V. Ayodele[1,2,3]* and Sureena Abdullah[1,3]*

*Address all correspondence to: bamidele.ayodele@uniben.edu; sureena@ump.edu.my

1 Faculty of Chemical and Natural Resources Engineering, Universiti Malaysia Pahang, Malaysia

2 Department of Chemical Engineering, University of Benin, Benin City, Nigeria

3 Center of Excellence for Advanced Research in Fluid Flow, Lebuhraya Tun Razak, Gambang Kuantan, Pahang, Malaysia

References

[1] Fawole OG, Cai X-M, MacKenzie AR. Gas flaring and resultant air pollution: A review focusing on black carbon. Environmental Pollution. 2016;**216**:182-197. DOI: 10.1016/j.envpol.2016.05.075

[2] James OO, Maity S, Mesubi MA, Ogunniran KO, Siyanbola TO, Sahu S, et al. Towards reforming technologies for production of hydrogen exclusively from renewable resources. Green Chemistry. 2011;**13**:2272. DOI: 10.1039/c0gc00924e

[3] Ayodele BV, Khan MR, Cheng CK. Syngas production from CO_2 reforming of methane over ceria supported cobalt catalyst: Effects of reactants partial pressure. Journal of Natural Gas Science and Engineering. 2015;**27**:1016-1023. DOI: 10.1016/j.jngse.2015.09.049

[4] Er H, Bouallou C, Werkoff F. Dry reforming of methane – Review of feasibility studies. Chemical Engineering Transactions. 2012;**29**:163-168. DOI: 10.3303/CET1229028

[5] Tsoukalou A, Imtiaz Q, Kim SM, Abdala PM, Yoon S, Müller CR. Dry-reforming of methane over bimetallic Ni–M/La_2O_3 (M=co, Fe): The effect of the rate of $La_2O_2CO_3$ formation and phase stability on the catalytic activity and stability. Journal of Catalysis. 2016. DOI: 10.1016/j.jcat.2016.03.018

[6] Ayodele BV, Khan MR, Cheng CK. Greenhouse gases abatement by catalytic dry reforming of methane to syngas over samarium oxide-supported cobalt catalyst. International Journal of Environmental Science and Technology. 2017;**14**(12):2769-2782. DOI: 10.1007/s13762-017-1359-2

[7] Ayodele BV, Khan MR, Cheng CK. Greenhouse gases mitigation by CO_2 reforming of methane to hydrogen-rich syngas using praseodymium oxide supported cobalt catalyst. Clean Technologies and Environmental Policy. 2017;**19**(3):795-807. DOI: 10.1007/s10098-016-1267-z

[8] Djinović P, Batista J, Pintar A. Efficient catalytic abatement of greenhouse gases: Methane reforming with CO_2 using a novel and thermally stable Rh-CeO_2 catalyst. International Journal of Hydrogen Energy. 2012;**37**:2699-2707. DOI: 10.1016/j.ijhydene.2011.10.107

[9] Yaw TC, Aishah NOR, Amin S. Analysis of carbon dioxide reforming of methane via thermodynamic equilibrium approach. Jurnal Teknologi. 2007;**43**:31-49

[10] Ayodele BV, Cheng CK. Process modelling, thermodynamic analysis and optimization of dry reforming, partial oxidation and auto-thermal methane reforming for hydrogen and syngas production. Chemical Product and Process Modeling. 2015;**10**:211-220. DOI: 10.1515/cppm-2015-0027

[11] Ginsburg JM, Piña J, El Solh T, De Lasa HI. Coke formation over a nickel catalyst under methane dry reforming conditions: Thermodynamic and kinetic models. Industrial and Engineering Chemistry Research. 2005;**44**:4846-4854. DOI: 10.1021/ie0496333

[12] Arbag H, Yasyerli S, Yasyerli N, Dogu T, Dogu G. Coke minimization in dry reforming of methane by Ni based mesoporous alumina catalysts synthesized following different routes: Effects of W and Mg. Topics in Catalysis. 2013;**56**:1695-1707. DOI: 10.1007/s11244-013-0105-3

[13] Bartholomew CH. Mechanisms of catalyst deactivation. Applied Catalysis A: General. 2001;**212**:17-60. DOI: 10.1016/S0926-860X(00)00843-7

[14] Pakhare D, Spivey J. A review of dry (CO_2) reforming of methane over noble metal catalysts. Chemical Society Reviews. 2014;**43**:7813-7837. DOI: 10.1039/c3cs60395d

[15] Argyle M, Bartholomew C. Heterogeneous catalyst deactivation and regeneration: A review. Catalysts. 2015;**5**:145-269. DOI: 10.3390/catal5010145

[16] Budiman AW, Song S-H, Chang T-S, Shin C-H, Choi M-J. Dry reforming of methane over cobalt catalysts: A literature review of catalyst development. Catalysis Surveys from Asia. 2012;**16**:183-197. DOI: 10.1007/s10563-012-9143-2

[17] Zhang WD, Liu BS, Zhan YP, Tian YL. Syngas production via CO_2 reforming of methane over Sm_2O_3-La_2O_3-supported Ni catalyst. Industrial and Engineering Chemistry Research. 2009;**45**:7498-7504. DOI: 10.1021/ie9001298

[18] Kathiraser Y, Oemar U, Saw ET, Li Z, Kawi S. Kinetic and mechanistic aspects for CO_2 reforming of methane over Ni based catalysts. Chemical Engineering Journal. 2015; **278**:62-78

[19] Xu W, Si R, Senanayake SD, Llorca J, Idriss H, Stacchiola D, et al. In situ studies of CeO_2-supported Pt, Ru, and Pt – Ru alloy catalysts for the water – Gas shift reaction: Active phases and reaction intermediates. Journal of Catalysis. 2012;**291**:117-126. DOI: 10.1016/j.jcat.2012.04.013

[20] Zhou H, Wu H, Shen J, Yin A, Sun L, Yan C. Thermally stable Pt/CeO_2 hetero-nanocomposites with high catalytic activity. Journal of the American Chemical Society. 2010;**132**(10):4998-4999

[21] Itkulova SS, Zhunusova KS, Zakumbaeva G. CO_2 reforming of methane over Co-Pd/Al_2O_3 catalysts. Bulletin of the Korean Chemical Society. 2005;**26**:2017-2020

[22] Bezerra MA, Santelli RE, Oliveira EP, Villar LS, Escaleira LA. Response surface methodology (RSM) as a tool for optimization in analytical chemistry. Talanta. 2008;**76**:965-977. DOI: 10.1016/j.talanta.2008.05.019

[23] Lundstedt T, Seifert E, Abramo L, Thelin B, Nyström Å, Pettersen J, et al. Experimental design and optimization. Chemometrics and Intelligent Laboratory Systems. 1998;**42**: 3-40. DOI: 10.1016/S0169-7439(98)00065-3

[24] Elksibi I, Haddar W, Ben Ticha M, Gharbi R, Mhenni MF. Development and optimisation of a non conventional extraction process of natural dye from olive solid waste using response surface methodology (RSM). Food Chemistry. 2014;**161**:345-352. DOI: 10.1016/j.foodchem.2014.03.108

[25] Khuri AI, Mukhopadhyay S. Response surface methodology. Wiley Interdisciplinary Reviews: Computational Statistics. 2010;**2**:128-149. DOI: 10.1002/wics.73

[26] Baş D, Boyacı İH, Bas D, Boyaci IH, Baş D, Boyacı İH, et al. Modeling and optimization I: Usability of response surface methodology. Journal of Food Engineering. 2007;**78**:836-845. DOI: 10.1016/j.jfoodeng.2005.11.024

[27] Singh P, Shera SS, Banik J, Banik RM. Optimization of cultural conditions using response surface methodology versus artificial neural network and modeling of l-glutaminase

production by *Bacillus cereus* MTCC 1305. Bioresource Technology. 2013;**137**:261-269. DOI: 10.1016/j.biortech.2013.03.086

[28] Mourabet M, El Rhilassi A, El Boujaady H, Bennani-Ziatni M, Taitai A. Use of response surface methodology for optimization of fluoride adsorption in an aqueous solution by Brushite. Arabian Journal of Chemistry. 2017;**10**:S3292-S3302. DOI: 10.1016/j.arabjc.2013.12.028

[29] Ishmael UC, Shah SR, Palliah JV, Asras MFF, Ahmad SS, Ayodele Bamidele V. Statistical modeling and optimization of enzymatic pretreatment of empty fruit bunches with laccase enzyme. BioResources. 2016;**11**:5013-5032

[30] Fan M-S, Abdullah AZ, Bhatia S. Hydrogen production from carbon dioxide reforming of methane over Ni–Co/MgO–ZrO_2 catalyst: Process optimization. International Journal of Hydrogen Energy. 2011;**36**:4875-4886. DOI: 10.1016/j.ijhydene.2011.01.064

[31] Aishah NOR, Amin S, Zakaria ZY. Optimization of oxidative coupling of methane using response surface methodology. Jurnal Teknologi. 2003;**39**:35-51

[32] Aishah NOR, Amin S, Yusof KM, Isha R. Carbon dioxide reforming of methane to syngas: Modeling using response surface methodology and artificial neural network. Jurnal Teknologi. 2007;**43**:15-29

[33] Sidik SM, Triwahyono S, Jalil AA, Majid ZA, Salamun N, Talib NB, et al. CO_2 reforming of CH_4 over Ni-Co/MSN for syngas production: Role of Co as a binder and optimization using RSM. Chemical Engineering Journal. 2016;**295**:1-10. DOI: 10.1016/j.cej.2016.03.041

[34] Braga TP, Santos RC, Sales BM, da Silva BR, Pinheiro AN, Leite ER, Valentini A. CO_2 mitigation by carbon nanotube formation during dry reforming of methane analyzed by factorial design combined with response surface methodology. Chinese Journal of Catalysis. 2014;**35**:514-523. DOI: 10.1016/S1872-2067(14)60018-8

[35] Hafizi A, Rahimpour MR, Hassanajili S. Hydrogen production by chemical looping steam reforming of methane over Mg promoted iron oxygen carrier: Optimization using design of experiments. Journal of the Taiwan Institute of Chemical Engineers. 2016. DOI: 10.1016/j.jtice.2016.01.023

[36] Mahmodi G, Sharifnia S, Rahimpour F, Hosseini SN. Photocatalytic conversion of CO_2 and CH_4 using ZnO coated mesh: Effect of operational parameters and optimization. Solar Energy Materials & Solar Cells. 2013;**111**:31-40. DOI: 10.1016/j.solmat.2012.12.017

[37] Ayodele BV, Khan MR, Nooruddin SS, Cheng CK. Modelling and optimization of syngas production by methane dry reforming over samarium oxide supported cobalt catalyst: Response surface methodology and artificial neural networks approach. Clean Technologies and Environmental Policy. 2016. DOI: 10.1007/s10098-016-1318-5

[38] Ayodele BV, Cheng CK. Modelling and optimization of syngas production from methane dry reforming over ceria-supported cobalt catalyst using artificial neural networks and Box-Behnken design. Journal of Industrial and Engineering Chemistry. 2015;**32**:246-258. DOI: 10.1016/j.jiec.2015.08.021

Chapter 6

Evaluation of Factors Affecting Chemical Extraction of Co Ions from Contaminated Soil

Ivana Smičiklas

Additional information is available at the end of the chapter

http://dx.doi.org/10.5772/68066

Abstract

Excessive concentrations of cobalt (Co) ions in the soil cause quality degradation and pose a significant hazard to biota. One of the options for the permanent separation of the pollutant from soil matrix is extraction by chemical reagents. In this study, response-surface methodology (RSM) was applied to evaluate the factors affecting Co extraction from contaminated calcareous soil. Solutions of disodium ethylenediaminetetraacetate (Na_2EDTA), citric acid (CA), and HCl were considered as leaching media. Reagent concentration, soil to solution ratio, and extraction time were selected as process variable, while Co extraction efficiencies and final pH values of extracts were the measured responses. The effect of factor variation between three levels was studied using Box-Behnken experimental design. By statistical analysis, the most influential factors were determined for each reagent, and the model equations were proposed for the prediction of system responses. Overlaid contour plots were used for the analysis of the effect of process conditions on both responses simultaneously. Given that each case of contamination is unique and requires extensive research before the remediation is implemented full-scale, it was shown that experimental design methodology is a smart approach for the assessment and comparison between the treatments.

Keywords: cobalt ions, soil contamination, chemical extraction, Box-Behnken design, ANOVA

1. Introduction

Cobalt (Co) is one of the essential heavy metals, which is vital at trace levels for proper functioning of the human metabolism [1]. However, at concentrations higher than optimal, essential metals become toxic. The poisoning by Co is commonly a result of drinking water contamination, high ambient air and soil concentrations, and consequent entrance and bioaccumulation of

Co in the food chain. Therefore, development and application of remediation methods are necessary in order to mitigate the negative impact of Co on the ecosystem.

The average concentration of Co in the Earth's crust is 10 mg/kg [2]. The levels of Co in the soil are influenced by the pedogenic factors, as well as by the characteristics of a parent rock material. For different climatic zones, the values of 0.05 and 300 mg/kg were reported as the minimum and maximum naturally occurring concentrations in the soil [3]. Lower amounts were found in the soils from the northern regions, which originated from the glacial deposits. The average concentrations of Co in the northern parts of Ukraine, Russia [2], and Sweden [4] are 3.5, 5.5, and 7.1 mg/kg, respectively. In contrast, naturally higher values can be found in the areas with arid or semiarid climate, like in Egypt [3].

The property which affects the manifestation of both beneficial and toxic effects is the solubility of Co in a particular environment. In respect to total concentration of Co in the soil, the fraction available to plants is more important. The accessibility to living organisms is the highest for free Co ions and water soluble complexes, while its metallic form and insoluble compounds commonly exhibit very low bioaccessibility. The available content of Co in different soil types, determined using 2.5% acetic acid as an extracting agent, was found to vary from 0.05 to 1 ppm [3]. On one hand, the lack of available Co leads to Co-deficiency in living organisms, while on the other, Co and its compounds are highly toxic in excessive amounts causing serious cell and tissue damages (LD_{50} values for intake of Co salts for rats are in the range 150–500 mg/kg of body weight [5]).

The main sources of environmental pollution with Co are industrial activities. Utilization of Co as catalyst in the chemical industry, in the production of dyes and pigments, magnetic recording media, alloys, batteries, etc., makes this element of the strategic importance for military, industrial, and commercial applications [6]. In numerous studies, Co concentrations in soil and sediments were measured to define its levels and identify the main pollution sources. Soil pollution by Co in Shenzhen soil (China) was attributed to uncontrolled discharge of industrial wastewater from factories that produce or use chemical compounds or alloys containing Co [7]. Furthermore, ceramic industry was highlighted as the source of the sediment contamination with Co and other heavy metals in Jiangsu Province (China) [8]. The highest concentrations were observed in the surface soil layer, within the 20–22 cm depth. In north greater Cairo (Egypt), direct discharge of industrial wastewaters to irrigation water canals during 30 years provoked significant contamination of soil with Co (146 mg/kg) [9]. The activity of a smelter in Zambia was found to result in dust with elevated concentrations of Co and other heavy metals, causing the contamination of soil and plants [10]. The highest measured Co concentration was 606 mg/kg found in the 0–2 cm of soil depth [11]. Furthermore, Co has been found in increased concentrations in at least 426, out of 1636 most serious hazardous waste sites in the USA, identified by Environmental Protection Agency (US EPA) [12]. The contamination of soil with Co is therefore a global problem with a tendency of steady increase.

Contaminated areas need to be treated using the techniques based either on the pollutant separation from the soil matrix or on its stabilization [13]. The levels and properties of heavy metal pollutants, as well as the physicochemical properties of the soil, govern the distribution of heavy metals which is an important factor for the selection of effective remediation method.

When soil is exposed to Co-containing solution, several processes can occur at the surface of soil constituents: electrostatic attraction, ion-exchange, complexation, co-precipitation, precipitation, occlusion, diffusion, and migration. Various single-step leaching procedures and sequential extractions are in use for the determination of metal distribution and mobility in the soil. Sequential extraction of the soil by so-called Tessier protocol is frequently applied [14]. This procedure is based on the consecutive application of selective chemical reagents in five extraction steps (F1–F5), with the aim to extract the pollutant associated with different soil fractions: ion-exchangeable (F1), acid soluble (F2), bound to Fe, Mn-oxides (F3), complexed by organic matter (F4), and incorporated in the residual fraction (F5). As the strength of the bonds between the metal and the soil constituents increases along the extraction scheme, the metal mobility decreases in the same manner.

Clean-up of the contaminated soil can be completed by the method of chemical extraction if the pollutant is distributed between relatively mobile fractions F1–F3 [15]. Extraction processes require mixing of contaminated soil with leaching solutions (solutions of acids, inorganic salts, chelating agents, surfactants, etc.) that cause a transfer of pollutant from the soil matrix into the liquid phase. If the chemical extraction is chosen among other alternative methods, the optimization of the method performance requires extensive research on the effects of a large number of variables. Some of the previous studies have addressed the impact of reagent type and concentration, reaction time, soil/solution ratio, pH, temperature, etc., on the efficiency of heavy metals extraction from the soil matrix [16–19]; however, the experiments were conducted by varying one factor at the time and little attention has been paid onto evaluation and comparison of factor effects, their interactions and process optimization.

Taking into account numerous potentially important factors, experimental design methodology (DOE) can be a useful strategy for the analysis of different soil treatments. In that sense, the present chapter aimed to explore the applicability of DOE approach for the analysis of the factors affecting chemical extraction of Co from contaminated soil and for the prediction of system responses.

2. Design of experiment (DOE)

The analysis and optimization of virtually any process can be conducted by experimental design methodology (DOE). Compared to the classical approach which implies variation of one factor at the time, DOE relies on simultaneous variation of all factors and reveals the most influential ones, the significant interactions between the factors, and the optimal levels of the factors. This methodology aims to describe or explain the variation of information under conditions that are hypothesized to reflect the variation. Therefore, DOE is a planned approach to find out the relationships between process variables and process responses, in relatively small number of experimental trials. Depending on the process under consideration, as well as on the number and the type of information of interest, different types of experimental design are in use, such as full and fractional factorial designs, response-surface designs, mixture designs, random block, Latine squares designs, etc. [20].

The response-surface methodology (RSM) was developed in order to optimize chemical reactions used on an industrial scale [21]. The application of RSM is of relevance in situations where several input variables influence the performance or quality characteristics of the process, which are called the responses. The input variables, usually denoted as independent variables, are controlled. Variation in experimental conditions provokes changes in the process response. Dependence between process responses and two process variables is a smooth surface called the response surface. The purpose of the response surface is to locate the conditions which lead to the achievement of the minimum/maximum response. Furthermore, in cases when several responses are of interest, RSM can give the optimal conditions for their target values [22]. The response values (y) are related to the process variables (x_1, x_2...) according to the model:

$$y = f(x_1, x_2, \ldots) + \varepsilon \tag{1}$$

In Eq. (1), $f(x_1, x_2 \ldots)$ is called the response surface and represents the mean response at each x, while ε represents other sources of variability (measurement error, background noise, the effect of other variables, etc.), and it is usually treated as a statistical error.

The function $f(x_1, x_2 \ldots)$ can be given as the first-order [Eq. (2)], or the second-order model [Eq. (3)]. The first-order model describes experimental conditions which provide the responses without peaks, and it is applicable for the description of local area responses without function extremes.

$$f(x_i) = \beta_0 + \sum \beta_i x_i + \varepsilon \tag{2}$$

In Eq. (2), β_i denotes the linear effect of the process variable x_i. Given that it includes only the main effects of the two or more variables, the form of the first-order model given by Eq. (2) is called the main effects model. In some processes, the interactions between studied variables may be significant; thus, the interaction term must be added to the first-order model, which introduce curvature into the response function.

The advantage of RSM is that the lack of fit can be estimated and the adequacy of the model can be estimated. The lack of fit test provides the answer whether the proposed model adequately describes system response (H_0—null hypothesis) or not (H_A—alternative hypotheses). If the proposed model is suitable, there is no lack of fit. When the first-order polynomial regression model is not adequate, the applicability of the second-order model is analyzed:

$$f(x_i) = \beta_0 + \sum \beta_i x_i + \sum \beta_{ij} x_i x_j + \sum \beta_{ii} x_i^2 + \varepsilon \tag{3}$$

The term $\beta_{i,j}$ represents the interaction effect between two factors x_i and x_j, and $\beta_{i,i}$ is the quadratic effect of factor x_i. The meanings of the other terms in Eq. (3) are the same as aforementioned. The designs that correspond to the surface response methodology are Box-Behnken design, central composite design, and optimal designs.

The results of a literature survey showed that different types of experimental design were used for the investigation of Co sorption by potential sorbent materials such as apatite, zeolite, bauxite residue, and dried activated sludge [23–25]; however, DOE approach has not been applied, to this point, for the study of Co sorption or leaching from the soil matrix. In fact, just a few examples of RSM application for the analysis of a soil remediation process can be found [26, 27]. In this study, a three-level-three-factor Box-Behnken experimental design was selected for the evaluation of the effect of factors which influence the Co extraction from contaminated soil. Box-Behnken design is a three-level factorial design which considers factors at their low (–1) and high (+1) values, as well as at their arithmetic mean (0) [28]. This is the type of quadratic design which does not contain the factorial or fractional factorial designs but considers the midpoints of the edges of the process space and its center. In general, it is an alternative to full factorial design at three levels.

3. Materials and methods

3.1. Soil contamination and clean-up experiments

The soil was sampled from the site of Vinča Institute of Nuclear Sciences (Belgrade, Serbia), from the surface layer (0–20 cm). Prior to contamination with Co ions, soil was dried at room temperature, homogenized, grained, and sieved in order to separate the fraction of particles with the diameter <2 mm. The soil at this locality was characterized as weak alkaline ($pH_{H2O} = 8$, $pH_{KCl} = 7$), with $CaCO_3$ content of 5.4%, total organic content (TOC) of 2.1%, and cation exchange capacity (CEC) of 13 meq/100 g [29]. Mineralogical analysis revealed quartz, kyanite, and muscovite as main crystal phases, whereas mica, albite, kaolinite, and calcium, magnesium–carbonate were present in lower quantities [29].

In the first step, the soil was artificially contaminated by Co ions. The solution containing 0.0012 mol/L of $Co(NO_3)_2$(Merck, p.a.) was mixed with 100 g of dried soil at the solid/liquid ratio 1:20. After 24 h of equilibration, the liquid phase was separated from the soil by filtration and the contaminated sample was dried in the air atmosphere. The concentration of Co ions sorbed by the soil was calculated as the difference between its initial and the residual concentration.

Soil clean-up was conducted using the method of chemical extraction. Based on the high extraction efficiencies observed in the previous study [18], the solutions of disodium ethylenediaminetetraacetate (Na_2EDTA), citric acid (CA), and HCl were selected as extracting agents. The contaminated soil was mixed with the reagent solutions in the centrifuge tubes which were placed on the rotary shaker and agitated at the constant speed (10 rpm) at ambient temperature ($20 \pm 2°C$).

The effects of the variation of reagent concentration, solid/liquid ratio, and contact time, among three levels, were investigated. Experimental variables and their levels are given in **Table 1**. The matrix with the experimental conditions for simultaneous variation of process parameters (**Table 2**) was generated using Minitab Release Software 13.1.

Extraction of Co ions was performed using each of the selected reagents, according to the conditions given in **Table 2**. After specified contact times, the suspensions were centrifuged for 10 min at 10,000 rpm and the extracted concentrations of Co ions were measured in clear supernatants. The chemical extraction experiments were conducted in duplicate.

Independent factor	Symbol	Level 1 coded value (−1)	Level 2 coded value (0)	Level 3 coded value (−1)
Reagent concentration (mol/L)	A	0.0005	0.05025	0.1
Soil/solution ratio	B	5	15	25
Contact time (h)	C	1	3.5	6

Table 1. Experimental factors and their levels used in the Box-Behnken design.

Experimental run	Independent factor		
	A	B	C
1	1	1	0
2	−1	0	−1
3	0	1	−1
4	−1	0	1
5	0	0	0
6	1	−1	0
7	0	1	1
8	−1	−1	0
9	0	0	0
10	0	−1	1
11	1	0	−1
12	0	−1	−1
13	1	0	1
14	−1	1	0
15	0	0	0

Table 2. Combinations of factors and their levels according to Box-Behnken design matrix.

In the sorption and extraction experiments, Co concentrations were determined by Perkin Elmer 3100 atomic absorption spectrometer (AAS), using a wavelength of 252.1 nm. Standards for instrument calibration were prepared by diluting certified Perkin Elmer standard (1000 mg/L), and the calibration was repeated after every 10-sample measurement. The detection limit was 0.05 mg/L, whereas the deviations among five replicate measurements for each sample were lower than 3%.

The percentages of the extracted Co ions were used as the main response function of the process. In addition, pH values of the leaching solutions after the completion of the process (denoted as final pH) were measured using WTW InoLab pH-meter, and also considered as system response.

3.2. Statistical analysis

Mean values of measured parameters, obtained from duplicate extraction experiments, were used as system responses for data interpretation and statistical analysis. Experimental point designation, analysis of variance (ANOVA), fitting of regression polynomial models, and graphical presentations (ternary plots) were performed using the statistical software MINITAB Release 13.2. The statistical analysis was done at the confidence level $\alpha = 95\%$.

4. Results and discussion

4.1. System responses at different experimental conditions

After the contamination step, the total amount of Co ions sorbed by the investigated soil was found to be 1390 mg/kg. High sorption affinities of soils and soil components toward Co are well documented [24, 30–32]. The sorbed amount is in agreement with the results of a previous study in which the soil from the same location was contaminated with Co under the range of experimental conditions [18].

The variation of chemical reagents and other experimental conditions according to the Box-Behnken design resulted in variation of system responses (**Figure 1**).

The mean values of Co extraction efficiency were found to fluctuate in the range 10–72% for Na_2EDTA, 1–66% for CA, and 0–71% for HCl as extracting agent (**Figure 1a**). The results signify that, at certain levels of the considered factors, amounts of extracted Co can reach high values using tested reagents. It is evident that the change of experimental conditions differently affects the performance of tested reagents and that effect of HCl was the lowest when compared to other reagents under the same sets of experimental conditions.

The final pH values were also largely dependent on the type of reagent and process variables (**Figure 1b**). The initial pH values of extracting solutions at their lower, middle, and higher selected concentrations were: 4.4, 4.6, and 4.9 for Na_2EDTA, 2.0, 2.2, and 3.5 for CA, and 1.0, 1.3, and 3.3 for HCl. After the reaction of contaminated soil with Na_2EDTA solutions, pH values were in the range 4.4–8.8, while using solutions of CA and HCl, pH values varied between 2.8–7.8 and 1.6–7.6, respectively. The observed increase in pH, after interaction with the soil, can be attributed to the buffering capacity of the soil which mainly originates from its carbonate content [18].

The knowledge of the nature and the strength of the bonds established between added Co ions and soil constituents are important for interpretation and understanding the effect of different chemical reagents. By means of the sequential extraction analysis, the portions of Co bonded in different fractions of the investigated soil were previously analyzed [18]. Considering various contamination levels and aging times (1 h to 30 days), Co ions were found to be associated with Fe, Mn-oxide fraction (F3), carbonate/acid soluble (F2), and ion-exchangeable (F1) fraction ($F1 + F2 + F3 > 92\%$). For the comparison, the majority of naturally occurring Co in the soil from the Vinča locality was found in F3 and residual (F5) phase [29], indicating that sorbed

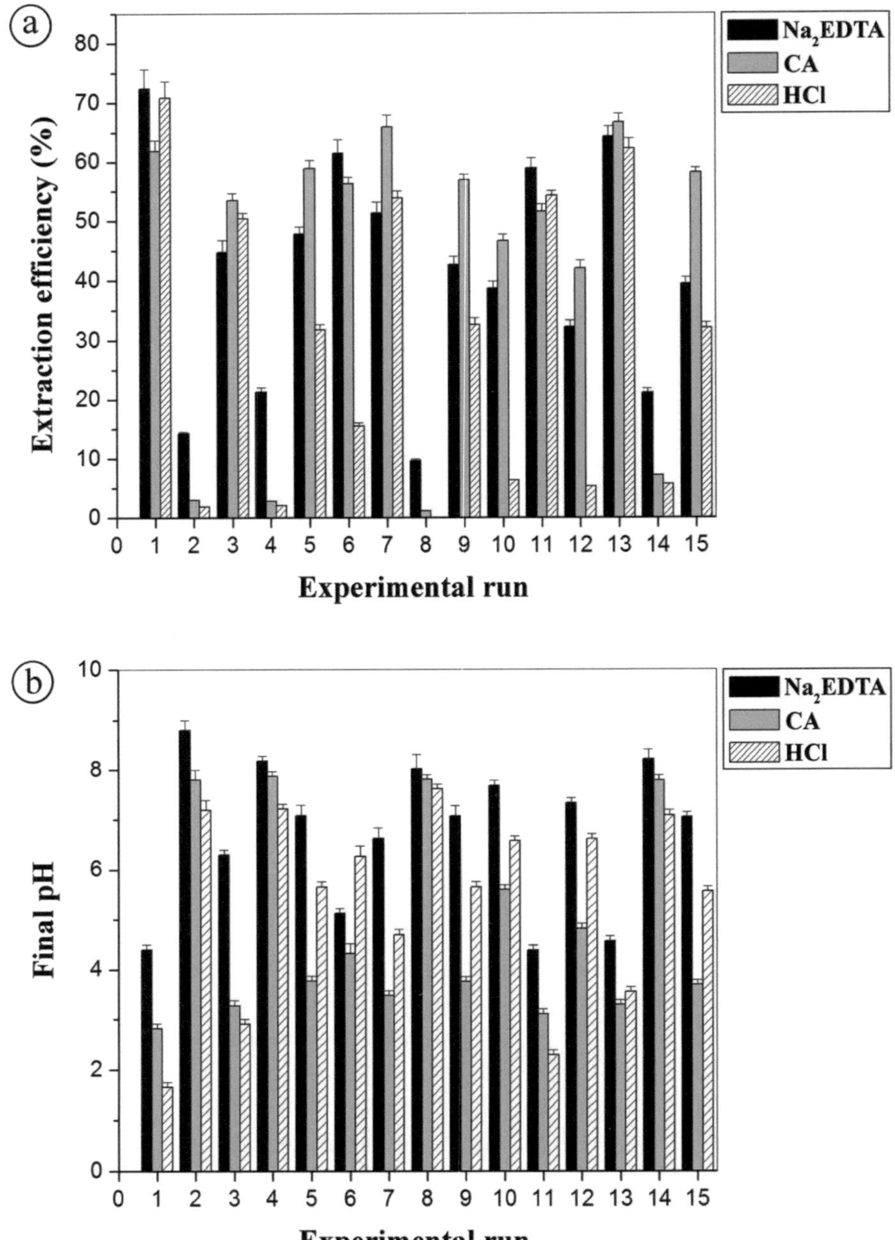

Figure 1. System responses obtained for different extracting agents under experimental conditions defined in **Table 2**: (a) extraction efficiency of Co ions from the contaminated soil and (b) final pH of extracts.

cobalt ions have significantly higher mobility. These results signify that chemical extraction can be considered for soil clean-up [33].

Fe- and Mn-oxides and hydroxides have been generally recognized as the major substrates for Co ions in the soil [34]. High affinities of such minerals toward Co ions were also revealed in the studies of Co sorption by Fe- and Mn-oxides [35–37], as well as by geochemical modeling of Co speciation in different soil samples [38]. The amounts of Co found in F2 phase of the sequential extraction correspond to the fraction associated with carbonates, as well as the fraction chemisorbed on the surface of soil components. The crystal lattice of calcium-carbonate can accommodate high quantities of Co and other divalent metals, implying that calcite is an important sorbent for metals in calcareous soils [39]. In addition, carbonate phase is of a crucial importance for high soil pH, which results in higher stability of sorbed metal ions [40]. Finally, the ion-exchange between Co and other divalent cations contained in the soil minerals also contributes to Co retention in the soil [41].

Taking into account the effects achieved by investigated reagents (**Figure 1a**), it can be concluded that both complexing agents and the mineral acid can release Co ions sorbed within different soil fractions. The extraction potential of HCl solutions is governed by the changes of pH. As the pH becomes more acidic, the dissolution of soluble soil constituents occurs, as well as the exchange reactions between H^+ and metal ions, and the protonation of active surface sites which become more positively charged [31, 42, 43]. Depending on the pH, these reactions cause the liberation of metal ions bonded by ion-exchange mechanism, specifically sorbed, and associated with carbonates and oxide minerals. Solubility of metal ions in the soil is also enhanced in the presence of chelating anions. These organic ions form multiple bonds with the metal cation, essentially in the form of a ring, which exhibits high stability in aqueous media. The chelating agents have a potential to extract the metal ions bonded to all non-residual fractions [44].

4.2. Evaluation of the effects of factors by Box-Behnken design

By using actual values of process variables, the response variable (y) was fitted to a second-order polynomial model as follows:

$$y = \beta_0 + \sum_{i=1}^{k} \beta_i x_i + \sum_{i=1}^{k} \beta_{ii} x_i^2 + \sum_{i=1}^{k-1} \sum_{j=i+1}^{k} \beta_{ij} x_i x_j \qquad (4)$$

where y is system response, β_0 is the intercept, β_i is the linear effect, and β_{ii} is the quadratic effect of the process variable x_i, while $\beta_{i,j}$ is the effect of the interaction between two independent process variables x_i and x_j.

Applicability of different models was tested by ANOVA. In order to determine which mathematical model adequately fits the obtained experimental results, full quadratic model was initially applied [Eq. (4)]. Statistical calculations were assessed on the basis of the F- and p-values. F-values are obtained from Fisher's test and represent the ratio of the mean square due to regression and mean square due to residual error, whereas p-value is defined as the smallest level of significance which leads to the rejection of null hypothesis. Therefore, higher F and lower p-values ($p < 0.05$) represent statistically significant terms. By omitting the insignificant terms ($p > 0.05$) and considering the values of determination coefficients (R^2) together with the results of lack of fit, the most appropriate models were found.

For Na_2EDTA as Co leaching reagent, the most suitable model included only linear terms, following the equation:

$$Co(Na_2EDTA) = 41.142 + 23.798A + 5.935B + 3.138C \tag{5}$$

According to the proposed Eq. (5), Co desorption efficiency was dependent on the linear effect of all selected factors. However, by observing the values of coefficients next to the respective factors, it can be concluded that the highest effect on the process response was achieved by the change of initial reagent concentration, followed by the effect of applied reagent volume, and finally the effect of contact time. The positive signs of the calculated coefficients indicate that leached amounts of Co increase with the increased levels of factors.

The coefficient of determination (R^2) was calculated according to the equation:

$$R^2 = 1 - \frac{SS_{res}}{SS_{tot}} \tag{6}$$

where SS_{res} denotes the sum of squares of the residuals, while SS_{tot} is the total sum of squares. SS_{tot} was calculated as the sum of squared differences between the different system responses values (y_i) and its mean. Furthermore, the SS_{res} is the sum of the squared differences between each system variable and its value predicted by the model. Knowing that the R^2 values increase with the number of terms in the model even if their effect is insignificant, an incorrect model could be proposed if R^2 magnitude is solely considered. Therefore, the value of the adjusted determination coefficient (R^2_{adj}) was included. In contrast to their effect on R^2 values, insignificant terms in the model provoke the decrease in R^2_{adj} values.

The summary of ANOVA, F and p-values obtained for significant model terms and the results of the lack of fit are given in **Table 3**. Using the experimental results obtained for Na_2EDTA extraction efficiency, calculated R^2 and R^2_{adj} values were 99.02 and 98.76%, respectively. Practically, the predicted model adequately describes even 99.02% of experimental results. Giving that F-value was <1 and p-value was >0.05, the proposed model can be regarded as adequate.

The following model was developed for CA as a leaching reagent:

$$Co(CA) = 54.527 + 27.779\ A + 5.304\ B + 3.990\ C - 23.238A^2 \tag{7}$$

The extraction of Co ions with CA solutions was found to be dependent on all investigated factors with the statistical significance. Changes in the level of reagent initial concentration have caused the highest impact on the leaching efficiency. Furthermore, the squared term in the derived equation (A^2) indicates the existence of curvature in the response surface. The obtained Eq. (7) fitted 97.88% of the experimental results, with the calculated R^2_{adj} value of 97.03%.

The extraction of Co using HCl can be described by the model which includes the following linear and interaction terms:

$$Co(HCl) = 28.31 + 24.15\ A + 19.20\ B + 12.39AB \tag{8}$$

For the proposed model, determined R^2 was 94.69%, while R^2_{adj} was 93.25%. The analysis of factors affecting the efficiency of applied inorganic acid indicates that the changes of contact

http://dx.doi.org/10.5772/68066

	Na_2EDTA		CA		HCl	
	F	*p*	*F*	*p*	*F*	*p*
Extraction efficiency						
Regression	371.70	<0.0001	115.50	<0.0001	65.45	<0.0001
A	1032.90	<0.0001	333.90	<0.0001	111.29	<0.0001
B	64.24	<0.0001	12.17	<0.006	70.40	<0.0001
C	17.96	0.001	6.89	0.025	/	/
A^2	/	/	109.04	<0.0001	/	/
AB	/	/	/	/	14.65	0.003
Lack of fit	4.48	0.196	5.37	0.160	3.13	0.310
Final pH of extracts						
Regression	86.21	<0.0001	96.28	<0.0001	77.93	<0.0001
A	318.98	<0.0001	219.89	<0.0001	269.36	<0.0001
B	9.91	0.01	18.17	0.001	132.38	<0.0001
A^2	13.48	0.004	50.79	<0.0001	/	/
C	/	/	/	/	10.61	0.012
C^2	/	/	/	/	8.95	0.017
AB	/	/	/	/	38.40	<0.0001
BC	/	/	/	/	7.86	0.023
Lack of fit	2.51	0.151	6.37	0.06	5.14	0.18

Table 3. Analysis of variance.

time did not significantly influence on the amount of leached cobalt. On the other hand, changes in the concentration and volume of the reagent were significant, as well as the interaction of these two parameters.

Furthermore, the models for the description of final pH values were developed. Considering the data obtained using Na_2EDTA, the following equation was found suitable, with calculated values of $R^2 = 96.48\%$ and $R^2_{adj} = 95.52\%$:

$$pH(Na_2EDTA) = 7.0186 - 1.8438A - 0.3250B - 0.5548A^2 \tag{9}$$

The final pH values of the CA extracts can be described ($R^2 = 96.33\%$, $R^2_{adj} = 95.33\%$) by Eq. (10):

$$pH(CA) = 4.0400 - 2.2175A - 0.6375B + 1.5600A^2 \tag{10}$$

Eqs. (9) and (10) indicate that variation of initial concentration and volume of both investigated complexing agents had an effect on final pH. The negative signs of the corresponding coefficients denote that increased concentration levels and increased levels of applied volume influenced the decrease in pH.

The complex mathematical model was obtained for describing the solution pH values after soil leaching with HCl:

$$pH(HCl) = 5.6386 - 1.9150A - 1.3425B + 0.3800C - 0.511C^2 - 1.0225AB + 0.4625BC \quad (11)$$

The adequacy of Eq. (11) was confirmed by high R^2 and R^2_{adj} values (98.32% and 97.06%, respectively). The model indicates that the variation of all studied factors, as well as AB and BC interactions, significantly affected the pH of resulting extracts. The HCl concentration and volume exhibited negative correlation with pH, while the contact time and pH were positively correlated.

Inside the range of experimental conditions covered by the design, the efficiency of Co extraction can be predicted using the mathematical models proposed by Eqs. (5), (7), and (8). The graphical interpretation of these equations, that help visualize the shape of the response surface, is given in **Figure 2** in form of contour plots. The graphs express the relationships between calculated

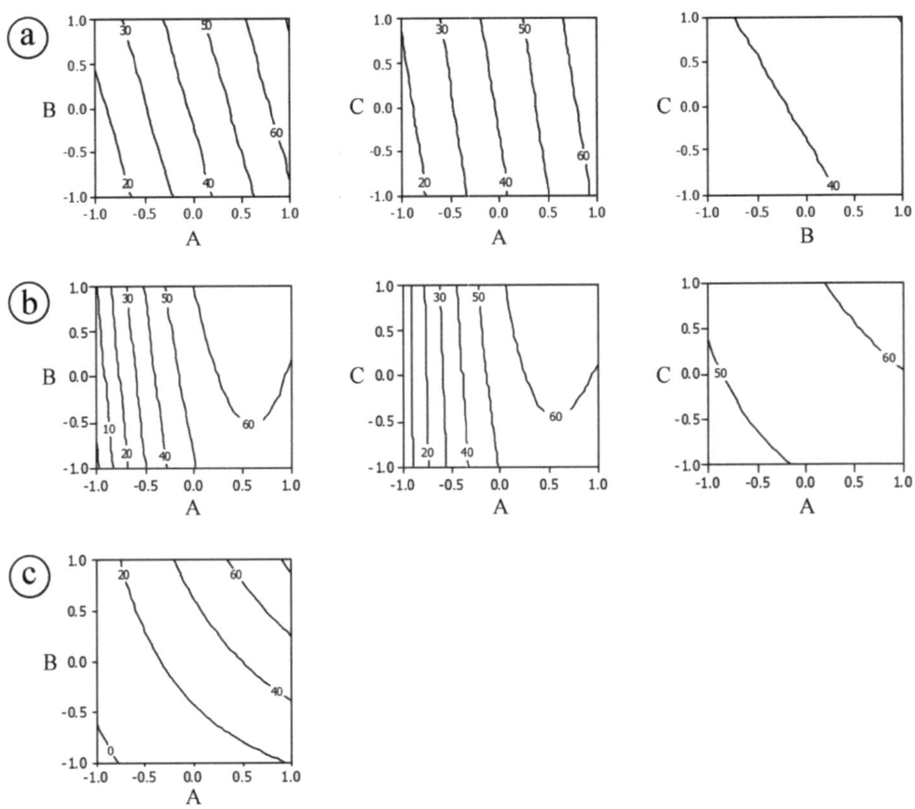

Figure 2. Contour plots for cobalt extraction efficiency using: (a) EDTA, (b) CA, and (c) HCl solutions, presenting the relationships between calculated responses and statistically significant factors, at the constant intermediate value of the third factor (A—reagent concentration, B—soil/solution ratio, C—extraction time).

http://dx.doi.org/10.5772/68066

responses and statistically significant variables, at the constant intermediate value of the third parameter. The lines on the graphs represent constant values of the system response.

The parallel lines on the contour plots in **Figure 2a** refer to the linear effect of process variables on Co desorption efficiency by Na_2EDTA, with the highest yield of extraction achieved at their maximum levels. The curvature in system responses occurred in case when CA was used as reagent, due to the quadratic term in the model equation, whereas in case of HCl due to the significant interaction effect between the reagent concentration and applied volume.

Similar predictions were made for final pH values as response functions, using Eqs. (9)–(11), and the contour plots are given for statistically significant parameters (**Figure 3**).

The contour graphs, as well as the ANOVA results, signify that increase of the initial reagent concentration provoked decrease in the final pH. Furthermore, pH declined with increased volume of the reagents. In contrast, longer reaction times contribute only to higher pH of HCl solution.

If multiple system responses are considered in the same experiment, their simultaneous analysis could be essential to optimize the process. Construction of overlaid contour plots enables determination of the ranges of process variable which lead to the achievement of the target effect. These plots are constructed by overlaying the contour plots of each considered response. The responses defined in the presented study, final pH values of the filtrate and extraction efficiency, are related: the lower the pH—the higher extraction yield (**Figures 2** and **3**). However, high acidity of extracting solutions may cause the degradation of soil vital properties due to significant degradation of both mineral and organic phase [43], making the soil inappropriate for on-site disposal and re-vegetation after completion of the treatment. Furthermore, the

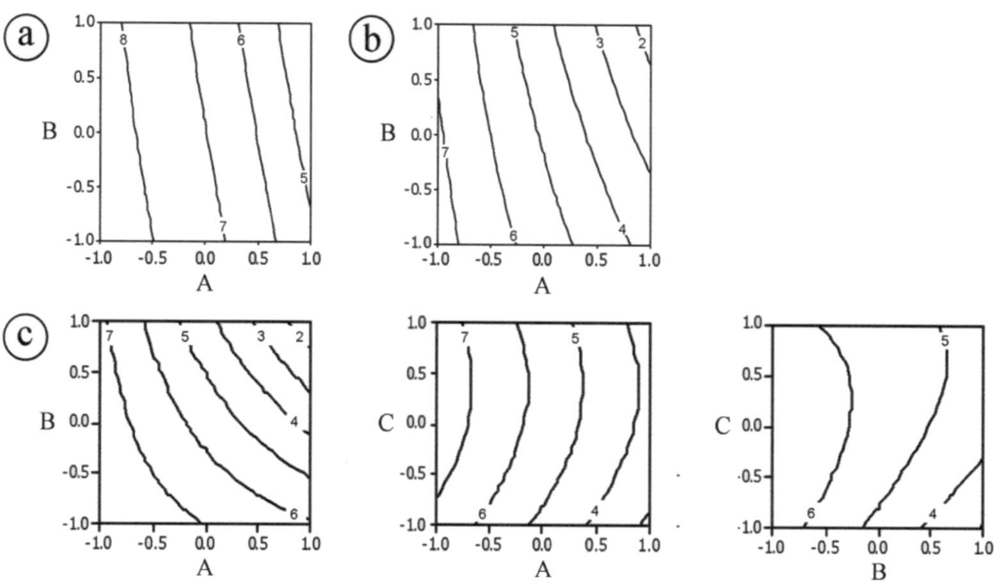

Figure 3. Contour plots for the final pH of extracts: (a) EDTA, (b) CA, and (c) HCl, presenting the relationships between calculated responses and statistically significant factors, at the constant intermediate value of the third factor (A—reagent concentration, B–soil/solution ratio, C—extraction time).

obtained filtrates represent liquid waste which requires further management (neutralization, metal recovery, reagent recovery, etc.). Thus, both the wastewater and the processed soil would need to be neutralized, increasing the complexity and cost of the treatment.

Therefore, the performance of different reagents was compared using near neutral final pH (5–7) and Co extraction efficiency >50%, as target values. Under these conditions, overlaid contour plots were constructed for Na_2EDTA and CA (**Figure 4a** and **b**). On the other hand, desired response values could not be reached simultaneously using HCL, and the target value of Co extraction was decreased to >40% (**Figure 4c**).

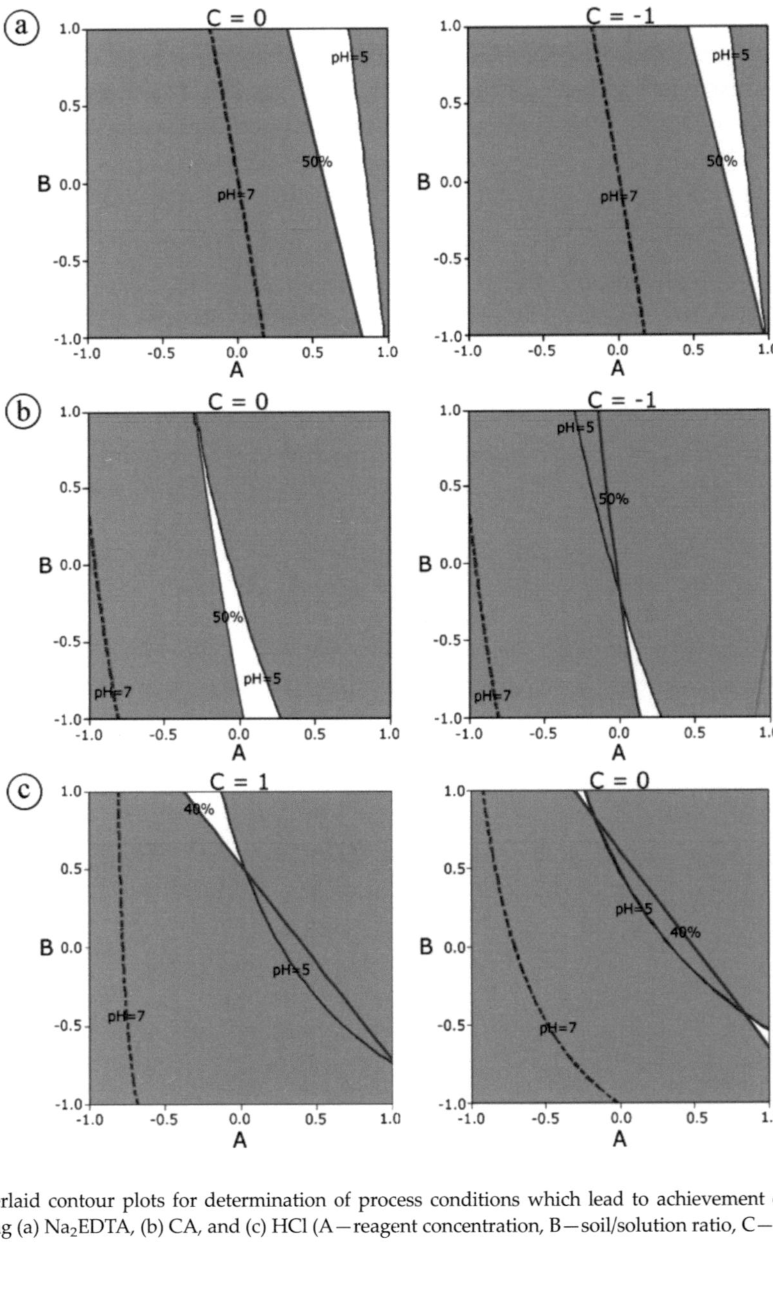

Figure 4. Overlaid contour plots for determination of process conditions which lead to achievement of target system responses using (a) Na_2EDTA, (b) CA, and (c) HCl (A—reagent concentration, B—soil/solution ratio, C—extraction time).

http://dx.doi.org/10.5772/68066

The white areas in the constructed plots present the ranges of factors A and B that provide given responses, at different contact times (C). Using Na_2EDTA, target response values can be obtained for contact times of 1 h, as well as 3.5 h, by selecting reagent concentration and volume in the designated ranges. Designated area is wider for 3.5 h, which practically means that if the duration of extraction is increased the desired responses can be gained at lower reagent concentrations. For the same extraction times, using CA instead of Na_2EDTA (**Figure 4b**), target values of Co desorption efficiency and solution pH can be obtained using initial concentrations and volumes of the reagent in much lower ranges. The effect of HCl was limited since it is governed by pH decrease (**Figure 4c**). Approximately 40% of sorbed Co ions can be leached under conditions that simultaneously assure target pH of the extract, at higher contact times (6 and 3.5 h).

Such results may be indicative for further evaluation of reagents. Consideration of the advantages and disadvantages related to the application of a chemical reagent includes the environmental impact and overall costs. Although HCl is often used for chemical leaching at full-scale, its effect is limited in calcareous soils. Consumption of high amounts of acid reagent would be necessary to provide efficient Co separation, which would in turn results in dissolution and degradation of soil matrix, creation of acidic waste water and soil residue. On the other hand, by selection of factor levels, complexing agents provide more efficient decontamination, even at near neutral pH conditions. Leaching of soil by chelating agents is potentially detrimental to its quality if a part of the chelating agent remains in the soil. As a natural compound, CA undergoes biodegradation process more easily compared to Na_2EDTA [45]. Furthermore, CA is less expensive chelating reagent [46], which supports the selection of CA among the tested reagents.

5. Conclusion

The Box-Behnken design was applied for screening and analyzing the factors affecting soil chemical extraction process using the sample of calcareous soil artificially contaminated with Co ions. Using Na_2EDTA, CA, and HCl as reagents, the effect of variation of three factors (concentration of the chemical reagent, soil/liquid ratio, and the contact time between the phases) between three levels was analyzed in respect to extraction efficiency and final pH values. The adequacy of different mathematical models, with inclusion of linear or quadratic terms, was tested for the description of experimental results. The extraction efficiency was highly dependent of the applied chemical reagent. Analysis of variance of the chosen responses revealed that Co separation was predominantly affected by the variation of the reagents concentrations. Effect of applied reagents volume had smaller statistical significance, while the contact time played an important role in the performance of complexing agents. Mathematical models were developed to describe the effect of each independent parameter and their interactions on system response. Predicted values of Co recovery, obtained using model equations, were in good agreement with the experimental data. Treatments by different reagents were compared using overlaid contour plots, taking into consideration extraction efficiencies in near neutral range of final pH values. The results demonstrate limited effect of HCl in cancerous soil, while chelating agents can exhibit high efficiencies by selection of factor levels. CA stands out as the most suitable agent, due to its performance, price, and biodegradability. The applicability of RSM for fast assessment and comparison of soil treatments was verified, highlighting the significance of DOE in practical applications.

Acknowledgements

This work was supported by the Ministry of Education, Science and Technological Development of the Republic of Serbia (Project III43009).

Author details

Ivana Smičiklas

Address all correspondence to: ivanat@vin.bg.ac.rs

Vinča Institute of Nuclear Sciences, University of Belgrade, Belgrade, Serbia

References

[1] Yamada K. Cobalt: Its role in health and disease. In: Sigel A, Sigel H, Sigel RKO, editors. Interrelations Between Essential Metal Ions and Human Diseases (Metal Ions in Life Sciences). Vol. 13. Springer, Netherlands; 2013, pp. 295-320. DOI: 10.1007/987-94-007-7500-8_9

[2] Kabata-Pendias A. Trace Elements in Soils and Plants. 4th ed. Boca Raton: CRC Press; 2010. p. 548. DOI: 10.1017/S0014479711000743

[3] Aubert H, Pinta M, editors. Development in Soil Science (Trace Elements in Soil). Vol. 7, Amsterdam: Elsevier; 1977. p. 394

[4] Eriksson J. Concentrations of 61 trace elements in sewage sludge, farmyard manure, mineral fertiliser, precipitation and in oil and crops. Report. Swedish Environmental Protection Agency, Stockholm, Sweden; 2001. Available from: http://swedishepa.se/Documents/publikationer/620-6246-8.pdf. [Accessed: December 20, 2016]

[5] Speijers GJA, Krajnc EI, Berkvens JM, Van Logten MJ. Acute oral toxicity of inorganic cobalt compounds in rats. Food and Chemical Toxicology. 1982;**20**:311-314. DOI:10.1016/S0278-6915(82)80298-6

[6] Donaldson JD, Beyersmann D. Cobalt and cobalt compounds. In: Ullmann's Encyclopedia of Industrial Chemistry, Weinheim: Wiley-VCH Verlag GmbH & Co. KGaA; 2005. DOI: 10.1002/14356007.a07_281.pub2

[7] Xu S, Lin C, Qiu P, Song Y, Yang W, Xu G, Feng X, Yang Q, Yang X, Niu A. Tungsten and cobalt-dominated heavy metal contamination of mangrove sediments in Shenzhen, China. Marine Pollution Bulletin. 2015;**100**:562-566. DOI: 10.1016/j.marpolbul.2015.08.031

[8] Liao QL, Liu C, Wu HY, Jin Y, Hua M, Zhu BW, Chen K, Huang L. Association of soil cadmium contamination with ceramic industry: A case study in a Chinese town. Science of the Total Environment. 2015;**514**:26-32. DOI:10.1016/j.scitotenv.2015.01.084

[9] Lotfy SM, Mostafa AZ. Phytoremediation of contaminated soil with cobalt and chromium. Journal of Geochemical Exploration. 2014;**144**:367-373. DOI: 10.1016/j.gexplo.2013.07.003

[10] Křibek B, Majer V, Knésl I, Nyambe I, Mihaljevič M, Ettler V, Sracek O. Concentrations of arsenic, copper, cobalt, lead and zinc in cassava (Manihot esculenta Crantz) growing on uncontaminated and contaminated soils of the Zambian Copperbelt. Journal of African Earth Sciences. 2014;**99**:713-723. DOI: 10.1016/j.jafrearsci.2014.02.009

[11] Ettler V, Mihaljevič M, Křibek B, Majer V, Šebek O. Tracing the spatial distribution and mobility of metal/metalloid contaminants in Oxisols in the vicinity of the Nkana copper smelter, Copperbelt province, Zambia. Geoderma. 2011;**164**:73-84. DOI: 10.1016/j.geoderma.2011.05.014

[12] Agency for Toxic Substances and Disease Registry (ATSDR), "Toxicological Profile for Cobalt," Atlanta, GA: US; 2004, Available from: https://www.atsdr.cdc.gov/toxprofiles/tp33.pdf [Accessed: 2016-12-15]

[13] Yao Z, Li J, Xie H, Yu C. Review on remediation technologies of soil contaminated by heavy metals. Procedia Environmental Sciences. 2012;**16**:722-729. DOI: 10.1016/j.proenv.2012.10.099

[14] Tessier A, Campbell PGC, Bisson M. Sequential extraction procedure for the speciation of particulate trace metals. Analytical Chemistry. 1979;**51**:844-851. DOI: 10.1021/ac50043a017

[15] Mulligan CN, Yong RN, Gibbs BF. Remediation technologies for metal-contaminated soils and groundwater: An evaluation. Engineering Geology. 2001;**60**:193-207. DOI: 10.1016/S0013-7952(00)00101-0

[16] Zou Z, Qiu R, Zhang W, Dong H, Zhao Z, Zhang T, Wei X, Cai X. The study of operating variables in soil washing with EDTA. Environmental Pollution. 2009;**157**:229-236. DOI: 10.1016/j.envpol.2008.07.009

[17] Yoo JC, Lee C, Lee JS, Baek K. Simultaneous application of chemical oxidation and extraction processes is effective at remediating soil Co-contaminated with petroleum and heavy metals. Journal of Environmental Management. 2017;**186**:314-319. DOI:10.1016/j.jenvman.2016.03.016

[18] Smičiklas I, Dimović S, Jović M, Milenković A, Šljivić-Ivanović M. Evaluation study of cobalt(II) and strontium(II) sorption–desorption behavior for selection of soil remediation technology. International Journal of Environmental Science and Technology. 2015;**12**:3853-3862. DOI: 10.1007/s13762-015-0817-y

[19] Naghipour D, Gharibi H, Taghavi K, Jaafari J. Influence of EDTA and NTA on heavy metal extraction from sandy-loam contaminated soils. Journal of Environmental Chemical Engineering. 2016;**4**:3512-3518. DOI: 10.1016/j.jece.2016.07.034

[20] Lazić Ž, Design of Experiments in Chemical Engineering. Weinheim: Wiley-VCH Verlag GmbH&Co.; 2004. p. 610. ISBN 3-527-31142-4

[21] Box GEP, Wilson KB. On the experimental attainment of optimum conditions. Journal of the Royal Statistical Society. 1951;**13**:1-45. DOI: 10.1007/978-1-4612-4380-9_23

[22] Dean AM, Voss D. Design and analysis of experiments. Springer Texts in Statistics. Springer, New York; 1999; 768 p, DOI:10.1007/b97673

[23] Smičiklas I, Sljivić-Ivanović M. Evaluation of factors influencing Co^{2+} removal by calcinated bone sorbent using experimental design methodology. Journal of Environmental Science and Health, Part A. 2012;**47**:896-908. DOI: 10.1080/10934529.2012.665006

[24] Šljivić-Ivanović M, Smičiklas I, Dimović S, Jović M, Dojčinović B. Study of simultaneous radionuclide sorption by mixture design methodology. Industrial & Engineering Chemistry Research. 2015;**54**:11212-11221. DOI: 10.1021/acs.iecr.5b03448

[25] Frišták V, Remenárová L, Lesný J. Response surface methodology as optimization tool in study of competitive effect of Ca^{2+} and Mg^{2+} ions in sorption process of Co^{2+} by dried activated sludge. Journal of Microbiology, Biotechnology and Food Sciences. 2012;**1**:1235-1249

[26] MartínezÁlvarez LM, Lo Balbo A, Mac Cormack WP, Ruberto LAM. Bioremediation of a petroleum hydrocarbon-contaminated Antarctic soil: Optimization of a biostimulation strategy using response-surface methodology (RSM). Cold Regions Science and Technology. 2015;**119**:61-67. DOI:10.1016/j.coldregions.2015.07.005

[27] Li Y-L, Fang Z-X, You J. Application of Box-Behnken experimental design to optimize the extraction of insecticidal Cry1Ac from soil. Journal of Agricultural and Food Chemistry. 2013;**61**:1464-1470. DOI: 10.1021/jf304970g

[28] Box GEP, Behnken DW. Some new three level designs for the study of quantitative variables. Technometrics. 1960;**2**:455-475. DOI: 10.1080/00401706.1960.10489912

[29] Dimović S, Smičiklas I, Šljivić-Ivanović M, Dojčinović B. Speciation of ^{90}Sr and other metal cations in artificially contaminated soils: The influence of bone sorbent addition. Journal of Soils and Sediments. 2013;**13**:383-393. DOI: 10.1007/s11368-012-0633-7

[30] Smičiklas I, Dimović S, Plećaš I. Removal of Cs^{1+}, Sr^{2+} and Co^{2+} from aqueous solutions by adsorption on natural clinoptilolite. Applied Clay Science. 2007;**35**:139-144. DOI: 10.1016/j.clay.2006.08.004

[31] Brown G, Parks G. Sorption of trace elements onto mineral surfaces: Modern perspectives from spectroscopic studies, and comments on sorption in the marine environment. International Geology Review. 2001;**43**:963-1073. DOI: 10.1080/00206810109465060

[32] Xu N, Hochella MF, Brown GE, Parks GA. Co(II) sorption at the calcite-water interface: I. X-ray photoelectron spectroscopic study. Geochimica et Cosmochimica Acta. 1996;**60**: 2801-2815. DOI: 10.1016/0016-7037(96)00133-0

[33] USEPA. Selecting remediation techniques for contaminated sediment. Report EPA-823-B93-001. Washington, DC, Cincinnati; 1993

[34] Bradl HB. Adsorption of heavy metal ions on soils and soils constituents. Journal of Colloid and Interface Science. 2004;**277**:1-18. DOI: 10.1016/j.jcis.2004.04.005

[35] Milenković A, Smičiklas I, Bundaleski N, Teodoro OMND, Veljović Đ, Vukelić N. The role of different minerals from red mud assemblage in Co(II) sorption mechanism. Colloids and Surfaces A. 2016;**508**:8-20. DOI: 10.1016/j.colsurfa.2016.08.011

[36] Al Abdullah J, Al Lafi AG, Al Masri W, Amin Y, Alnama T. Adsorption of cesium, cobalt, and lead onto a synthetic nano manganese oxide: Behavior and mechanism. Water, Air, & Soil Pollution. 2016;**227**:241. DOI: 10.1007/s11270-016-2938-4

[37] Mukherjee J, Ramkumar J, Chandramouleeswaran S, Shukla R, Tyagi AK. Sorption characteristics of nano manganese oxide: Efficient sorbent for removal of metal ions from aqueous streams. Journal of Radioanalytical and Nuclear Chemistry. 2013;**297**:49-57. DOI: 10.1007/s10967-012-2393-7

[38] Pourret O, Lange B, Houben D, Colinet G, Shutcha M, Faucon M-P. Modeling of cobalt and copper speciation in metalliferous soils from Katanga (Democratic Republic of Congo). Journal of Geochemical Exploration. 2015;**149**:87-96. DOI: 10.1016/j.gexplo.2014.11.011

[39] Zachara JM, Cowan CE, Resch CT. Sorption of divalent metals on calcite. Geochimica et Cosmochimica Acta. 1991;**55**:1549-1562. DOI: 10.1016/0016-7037(91)90127-Q

[40] Rieuwerts JS, Thornton I, Farago ME, Ashmore MR. Factors influencing metal bioavailability in soils: Preliminary investigations for the development of a critical loads approach for metals. Chemical Speciation and Bioavailability. 1998;**10**:61-75. DOI: 10.3184/095422998782775835

[41] Harter RD. Competitive sorption of cobalt, copper, and nickel ions by a calcium-saturated soil. Soil Science Society of America Journal. 1992;**56**:444-449. DOI: 10.2136/sssaj1992.03615995005600020017x

[42] Kuo S, Lai MS, Lin CW. Influence of solution acidity and $CaCl_2$ concentration on the removal of heavy metals from metal-contaminated rice soils. Environmental Pollution. 2006;**144**:918-925. DOI: 10.1016/j.envpol.2006.02.001

[43] Dermont G, Bergeron M, Mercier G, Richer-Laflèche M. Soil washing for metal removal: A review of physical/chemical technologies and field application. Journal of Hazardous Materials. 2008;**152**:1-31. DOI: 10.1016/j.jhazmat.2007.10.043

[44] Wuana RA, Okieimen FE, Imborvungu JA. Removal of heavy metals from a contaminated soil using organic chelating acids. International Journal of Environmental Science and Technology. 2010;**7**:485-496. DOI: 10.1007/BF03326158

[45] Wen J, Stacey SP, McLaughlin MJ, Kirby JK. Biodegradation of rhamnolipid, EDTA and citric acid in cadmium and zinc contaminated soils. Soil Biology and Biochemistry. 2009;**41**:2214-2221. DOI: 10.1016/j.soilbio.2009.08.006

[46] Kim GN, Jung YH, Lee JJ, Moon JK, Jung CH. An analysis of a flushing effect on the electrokinetic-flushing removal of cobalt and cesium from a soil around decommissioning site. Journal of Hazardous Materials. 2008;**152**:1-31. DOI: 10.1016/j.jhazmat.2007.10.043

Chapter 7

Design of Experiment Approach in the Industrial Gas Carburizing Process

Muhammad Atiq Ur Rehman,
Muhammad Azeem Munawar, Qaisar Nawaz and
Muhammad Yousaf Anwar

Additional information is available at the end of the chapter

http://dx.doi.org/10.5772/intechopen.72822

Abstract

Carburized samples were prepared under different sets of conditions at Millat Equipment Limited, Lahore, Pakistan, using continuous carburizing furnace under a reducing atmosphere. The gas carburizing process parameters were determined by the Taguchi design of experiment (DoE), an orthogonal array of L9 type with the mixed level of control factors. The key process parameters in gas carburizing process such as delay quenching interval, hardening temperature, and soaking time in oil were optimized in terms of core hardness, effective case depth (ECD), and surface hardness. DoE approach elucidated that the best results in terms of core hardness are A2 (delay quenching for 60 seconds), B2 (hardening temperature of 800°C), and C2 (soaking in quenching oil for 300 seconds). However, the best results in terms of ECD were A1 (delay quenching for 45 seconds), B3 (hardening temperature of 820°C), and C1 (soaking in quenching oil for 180 seconds). In order to choose the optimized parameters from the results given by DoE, microscopic analysis was conducted. Microscopic analysis showed coarse bainitic structure in core and tempered martensite at the surface of the samples processed at A2 (delay quenching for 60 seconds), B2 (hardening temperature of 800°C), and C1 (soaking in quenching oil for 180 seconds) compared to the other process conditions (A1, B3, and C1), which shows fine bainitic structure at core and relatively higher amount of retained austenite at the surface. Finally, defect per million opportunities (DPMO) model exhibited that the samples produced from the optimized set of parameters (A2, B2, and C1) are highly reproducible, gaining DPMO of 83 parts per million (PPM).

Keywords: gas carburizing, core hardness, design of experiment, defect per million opportunities, effective case depth

1. Introduction

In the current era, most of the industries are focusing on economical process optimization techniques. The most common industrial approaches are trial-and-error approach or one variable at a time (OVAT) in which one variable is changed at one time [1, 2]. These approaches are inefficient and time-consuming. However, statistical tools, such as the Taguchi design of experiments (DoE), provide superior methods to conduct experiments by efficient means. The selection of investigated parameters involved in the process is based on the philosophy of Dorian Shainin, which is characterized by focusing on a limited number of parameters selected by their cause and effect relationship [2, 3]. Hence, the Taguchi DoE has been widely used nowadays in the industrial sector. Moreover, Taguchi DoE approach allows changing more than one factor at one time, which reduces the number of experiments required to determine the optimized parameter [2, 4–6].

The Taguchi DoE assesses the effect and significance of controllable factors of an experiment, which increases the robustness of a process. Robustness is measured as a signal-to-noise (S/N) ratio. It comprises the sensitivity of the signal to disturbing factors involved in the process, the so-called noise [4, 7]. The optimization process is based on the assessment of this ratio as it determines the impact of control factors on the process. The DoE provides an efficient method for the optimization since only a limited number of experiments are required in contrast to full factorial methods [3, 8–10]. The statistical significance of each control factor can be determined by the performance of the multivariate analysis of variance (MANOVA). The significance of individual control factors can be estimated by its probability value p [5, 7].

Another important factor in the industrial process is the reproducibility. Design of experiment approach can further help in improving the reproducibility of the process. An important method to determine the reproducibility of the process is defect per million opportunities model (DPMO). DPMO evaluates the reproducibility of the process to relatively high sensitivity such as parts per million (PPM) [11, 12]. The "Six Sigma" approach can help in achieving the highly reproducibility of the industrial process. Therefore, it can be concluded that DoE approach coupled with the Six Sigma can help in economical process optimization at higher statistical confidence and reproducibility. Finally, KAIZEN "tools for continuous improvement" can help in sustaining the optimized experimental conditions [12].

Taguchi DoE is applied here to solve the particular industrial problems related with the higher field failure rate of crown wheel pinion. For example, field failure for the pinion after only 200 working hours was taken into account. Microscopic examination of sectioned piece revealed Chevron nature of fracture. Metallographic studies were also taken into account to know the reason behind this premature failure. Findings were the higher value of core hardness (38 HRC), with 25% of retained austenite. In service the retained austenite, which is metastable phase, will transform into martensite raising brittleness, because newly formed martensite is untempered. In addition, this newly formed martensite will cause dimensional changes as well as cause unexpected shift of contact pattern and backlash of more than 0.01 mm. Higher value of core hardness will make the core to respond to any sudden shock with minimum absorptivity and maximum transmissibility [13–15]. Furthermore, literature was studied to investigate the reasons behind the higher values of core hardness and retained austenite. It was found that high quenching temperatures, high quenching oil's temperature, improper diffusion, over carburizing, coarse initial grain size, segregation of the impurities at high-angle grain boundaries, higher value of chromium and silicon, improper soaking in the quenching oil, and viscosity of the oil might also be the reason [16–18]. Higher values of core hardness and retained austenite cause increase in the surface brittleness and improper load distribution behavior under dynamic loading conditions. It is hoped that this research will have paramount importance for industrial development and academic research [17, 19, 20]. Therefore, keeping in view for the criticality of core hardness, it was aimed to develop an intermediate and compromised value of core hardness during the carburizing process. In this research work, we choose delay quenching time, soaking time in oil, and hardening temperature as the significant parameters on the basis of fishbone diagram. Afterward, the effect of each parameter at various levels was studied to determine the optimum gas carburizing conditions.

2. Materials and methods

2.1. Materials

The material used in the present experimental work was a low-carbon low-alloy steel (SAE 8620). The chemical composition of the steel is given in **Table 1**.

The addition of alloying elements such as Mn, Mo, Ni, and Cr increases the hardening ability of the steel. The depth to which the steel is hardened is usually called as the hardenability of the steel.

2.2. Gas carburizing process

Test coupons were made up of 8620 low-alloy steel having dimensions of 2 × 1 (inch^2). One of the test samples was subjected to destructive testing for the metallography (microstructure

Element	Composition (mol %)
C	0.18–0.23
Si	0.15–0.35
Mn	0.7–0.9
Mo	0.150–0.25
S	0.04 max
P	0.35 max
Ni	0.4–0.7
Cr	0.4–0.6

Table 1. Chemical composition for 8620 alloy steel determined by using spark arc emission spectroscopy at AFCO Steel Mills Pvt Ltd., Lahore, Pakistan.

is shown in **Figure 5(D)**). Hardness was measured for all the samples prior to carburizing (150–200 HB, Brinell hardness number). Samples were charged into the continuous carburizing furnace (Gibbons furnace, UK) having fixed carburizing time (3.5 hours), carburizing temperature (930°C), and carbon potentials (1.00), but varying quenching time, holding in air (delay quenching intervals) before quenching, and hardening temperatures. It is important to highlight that carburizing time, carburizing temperature, and carbon potential were determined from the fishbone diagram (data not shown here), following previous studies [7, 10, 21]. Endo-gas (CO) was provided from endothermic gas generator to maintain reducing environment in the furnace. Diffusion of carbon took place in the same furnace but at comparatively low temperatures (780–840°C) and low values of flow rates of enrichment gas than in the carburizing process. Quenching was done in quenching tank of 12 × 10 × 8 feet at 75°C. Samples were tempered in the conveyer type tempering furnace at 120°C; prior to this washing was done.

2.3. Metallography

Metallography consists of studying the microscopic structure and characteristics of a given metal or an alloy. Metallography determines grain size, grain shape distribution of various phases and inclusions, and mechanical and thermal treatment of metals [22]. The non heat treated samples were sectioned using manual hacksaw whereas for cutting the heat treated samples abrasive cutoff wheel was used [22, 23].

Samples were cut and machined into 1 × 2 ($inch^2$) for the carburizing process, as shown in **Figure 1**. Test coupons after the gas carburizing process were subjected to the destructive testing such as effective case depth measurement and microstructural analysis. Prior to these samples were grind and polished (**Figure 2**). Two percent of nital was used to etch (to reveal the internal microstructure) the samples followed by washing with ethyl alcohol and subsequent

Figure 1. Optical image of the machine test coupon used for the carburizing process.

Figure 2. Digital image of the mounted samples after cutting, grinding, and polishing showing a mirror-like surface.

drying [22]. Etchant attacks the particles at high energy levels, since grain boundaries are at higher energy level so they become visible under microscope. The microstructure was evaluated at the surface and core with LECO microscope at MEL Quality Assurance Department (microstructures after heat treatment are shown in **Figure 5(A–C)**).

2.4. Effective case depth (ECD) measurement

ECD was measured with "Micro Vickers Hardness Testing Machine" (Shimadzu, Japan). Polished samples were placed on the platform, and then harness was measured by the indent steps at 0.1 mm. After taking an indent, mean diagonal length was measured and converted into Vickers pyramid hardness number (VPN). Similar indents were taken at increasing distance till the hardness value of 500 VPN. The distance from surface at which 500 VPN is achieved is known as effective case depth (ECD) [22]. Core hardness was also measured by taking an indent at the center point of test sample.

2.5. Hardness testing

After carburizing samples were subjected to surface hardness test using Rockwell hardness tester in which the first minor load (10 Kg_f) was applied followed by the major load of 300 Kg_f.

2.6. Design of experiments

In order to optimize the gas carburizing of SAE 8620 alloy steel, the basic tools of Six Sigma were applied. Vital parameters offered by the basic diffusion model was scrutinized by cause and effect diagram in terms of intended application (hard case and tough core), which offered the basis for selection of control factors (soaking time in oil, delay quenching interval, and hardening time), as shown in **Table 2**. In order to study the effect of these control factors, an orthogonal array of L9 (three control factors and three levels of each) type with the mixed level of control factors was applied (**Table 3**). The DoE approach was helpful in reducing the number of experiment from 27 (full factorial design) to 9 (DoE array of L9 type). Thus, the optimized parameters can be determined with the less number of experiments.

Symbol	Control factor	Levels		
		1	2	3
A	Delay quenching interval (S)	45	60	90
B	Hardening temperature (°C)	780	800	820
C	Soaking time in oil (S)	180	300	420

Table 2. Control factors and level of variables.

Run	Control factors			Core hardness (HRC)	ECD (mm)	Surface hardness (HRC)	S/N ratio for core hardness (dB)
	Delay quenching interval (S)	Hardening temperature (°C)	Soaking time in oil (S)				
1	45	780	180	26	0.90	57	28,299
2	45	800	300	29	1.00	59	29,248
3	45	820	420	34	1.20	60	30,629
4	60	780	300	22	0.65	58	26,848
5	60	800	420	25	0.80	57	27,958
6	60	820	180	28	0.90	58	28,943
7	90	780	420	20	0.45	56	26,020
8	90	800	180	23	0.60	57	27,234
9	90	820	300	26	0.70	57	28,299

Table 3. Experimentally measured values of core hardness, ECD, and surface hardness for gas carburizing process of SAE 8620 steel.

3. Results and discussion

3.1. Gas carburizing process

In the gas carburizing process, first the test coupons were preheated at 400°C for 30 minutes to avoid thermal shocks. Second, the temperature was raised to 930°C under a reducing environment, which was produced by the flow of CO in the furnace from endothermic gas generator. Moreover, methane gas was enriched in the furnace to maintain the carbon potential of 1.0. At 930°C the solubility of carbon in steel is approximately 1.14 (because steel transforms to the austenitic phase), due to which carbon started to flow from the atmosphere to the samples [18, 24, 25]. After soaking the samples for optimized time period, the surface carbon contents of the samples were raised. Afterward, samples entered into the hardening zone/diffusion zone, where the samples were kept for 1 hour in the temperature range of 780–820°C under a reducing environment [26, 27]. Since the temperature of the samples are dropped now, carbon from the outer surface starts to diffuse into the core [28]. The reason could be that outer surface is at lower temperature in comparison to that of inner surface of the sample; thus, the solubility of inner portion is reasonably high, which allows the diffusion of carbon [25, 29]. Finally, samples were quenched in the mineral oil at 75°C. The quenching process allows the diffusionless martensitic transformation at the surface [30–32]. However, thermal gradient at the core of the sample is significantly less, which allows the bainitic or ferritic transformation to occur in the core of the samples [32, 33]. The carburizing cycle of the test coupons is shown in **Figure 3**.

3.2. Design of experiment approach for gas carburizing of SAE 8620 steel

Control factors and their levels used in the design of experiment approach are illustrated in **Table 2**.

After the selection of control factors with the help of a cause and effect diagram (not given here), a signal-to-noise ratio (S/N) was calculated for the average core hardness of the heat-treated samples by considering that high value of S/N is desired [4]:

$$\frac{S}{N} = -10\,log\left[\frac{1}{n}\left(\frac{1}{\sum_i^2 Y}\right)\right] \qquad (1)$$

where

Y_i is the average core hardness.

n the number of observations.

Unit for S/N ratio is dB.

The design of experiment technique allows studying the effects of each parameter at different levels by averaging S/N ratio at each level. For example, the mean S/N ratio for deposition yield at levels 1, 2, and 3 of control factor A (delay quenching interval) can be calculated by

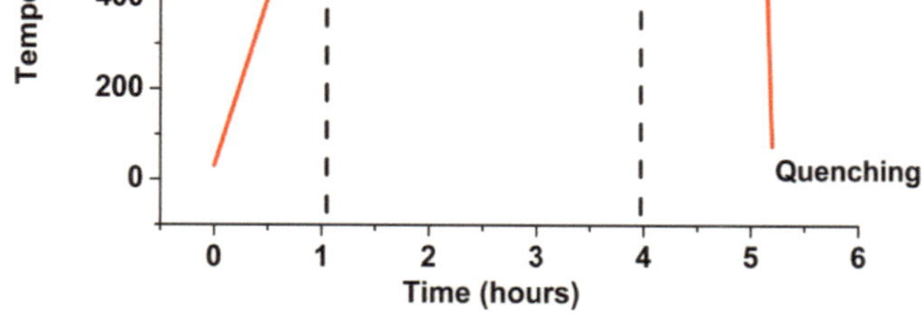

Figure 3. Typical carburizing cycle of the Gibbons furnace for SAE 8620 test coupons.

averaging S/N ratios for experiments 1–3, 4–6, and 7–9, respectively [4, 5, 7]. This technique was used for calculating the response of core hardness and ECD as presented in **Table 3**.

DoE approach was conducted to find out the suitable combination of gas carburizing parameters in order to ensure the optimum gas carburizing parameters with relatively higher statistical confidence. The effect of each control factor on the core hardness and ECD is shown in **Figure 4**, which shows that the parameters, that is, delay quenching interval and hardening temperature, have significant effect on core hardness and case depth. This effect was further confirmed by calculating maximum-minimum values reported in **Tables 4–7**.

Figure 4(A) shows that the increase in delay quenching significantly reduces the core hardness. The increase in hardening temperature tends to increase the core hardness. However, the increased soaking time in oil slightly raises the core hardness value. The aim was to obtain an intermediate value of core hardness; thus, the core should neither be too hard to cause fracture nor too soft to cause core crushing [10, 17, 34]. The intermediate values for the core hardness were attained at A2 (delay quenching interval of 60 seconds), B2 (hardening temperature of 800°C), and C2 (soaking time in oil for 300 seconds) conditions, as shown in **Table 8**. The reason for the decrease in the core hardness with the rise in delay quenching intervals could be the slow cooling of core in air compared to that of oil, which may have led to ferritic

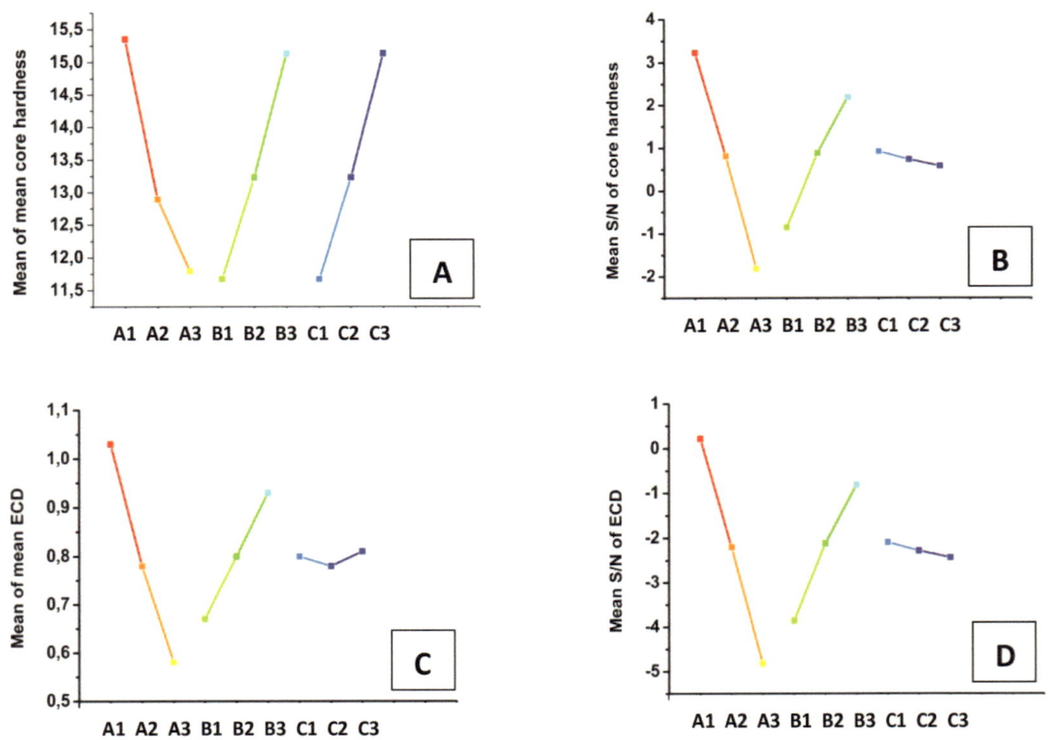

Figure 4. (A) Mean of mean core hardness response of control factors on core hardness, (B) mean of S/N (dB) response for the effect of control factors on core hardness, (C) mean of mean response for the effect of control factors on ECD, and (D) mean of S/N (dB) response for the effect of control factors on ECD.

Level	Delay quenching interval (S)	Hardening temperature (°C)	Soaking time in oil (S)
1	15.35	11.67	13.23
2	12.89	13.23	13.23
3	11.79	15.13	13.58
Maximum-minimum	3.56	3.47	0.35
Rank	1	2	3

Table 4. Mean response for the core hardness.

Level	Delay quenching interval (S)	Hardening temperature (°C)	Soaking time in oil (S)
1	3.23	−0.86	0.92
2	0.81	0.88	0.73
3	−1.82	2.20	0.58
Maximum-minimum	5.04	3.05	3.41
Rank	1	2	3

Table 5. S/N (dB) response for the core hardness.

Level	Delay quenching interval (S)	Hardening temperature (°C)	Soaking time in oil (S)
1	1.03	0.67	0.80
2	0.78	0.80	0.78
3	0.58	0.93	0.81
Maximum-minimum	0.45	0.27	0.03
Rank	1	2	3

Table 6. Mean response for ECD.

Level	Delay quenching interval (S)	Hardening temperature (°C)	Soaking time in oil (S)
1	0.22	−3.86	−2.09
2	−2.20	−2.12	−2.28
3	−4.82	−0.81	−2.43
Maximum-minimum	5.05	3.05	0.34
Rank	1	2	3

Table 7. S/N (dB) response for ECD.

	Delay quenching interval (S)	Hardening temperature (°C)	Soaking time in oil (S)
Optimum parameters for the best ECD	45 (A1)	820 (B3)	180 (C1)
Optimum parameters for the best core hardness	60 (A2)	800 (B2)	300 (C2)
Optimum parameters for the best combination of ECD, core hardness, and microstructure	60 (A2)	800 (B2)	180 (C1)

Table 8. Illustration of the best possible condition for optimum ECD, core hardness, and the combination of core hardness, ECD, and microstructure.

transformation at core rather than the bainitic transformation [21, 30], thus lowering the core hardness values of the specimens. Similar effect may be held responsible for the drop in core hardness values with the decrease in immersion time in oil [22].

Figure 4(C) shows the effect of control factors on the effective case depth. Here, the maximum value of case depth is required to provide high wear resistance for longer period of time [33]. The most significant parameter was the delay quenching interval. Reducing the delay time significantly improves the ECD, which may be due to the fact that rapid quenching hinders the diffusion of carbon toward the core of the sample. Thus, we formed uniform and shallow case depth during the hardening process. It was inferred from **Figure 4(C** and **D)** that best parameters in terms of the ECD are A1 (delay quenching of 45 seconds), B3 (hardening temperature of 820°C), and C1 (soaking in oil for 180 seconds).

It was concluded that there is a contrast in the best results for ECD and core hardness. Therefore, it was essential to establish a trade-off between the ECD and core hardness. Thus, we choose A2 (delay quenching of 60 seconds), B2 (hardening temperature of 800°C), and C1 (soaking in oil for 180 seconds), as the best condition. Soaking time in oil was the least significant factor; thus, it was chosen on the basis of the highest possible case depth. On the other hand, A2 and B2 were selected on the basis of core hardness because it was established in our previous studies that hard core will cause the premature failure of the components. Moreover, the microstructural analysis of the samples hardened at 820°C shows the formation of relatively higher amount of retained austenite (almost 25%, as shown in **Figure 5(A)**), which is a brittle phase and may also undergo dimensional changes, because retained austenite will eventually transform into the martensite in service [17, 19, 30]. It is important to highlight that the quantitative analysis was done with the software equipped with the LECO microscope, which quantitatively gives the amount of each phase present in the microstructure. Dimensional changes during the operation/service are deleterious, which may result in uneven load distribution between the matting parts. For example, the field failure investigations of different components show the deviations in pitch circle diameter (PCD), face runout (FR), and backlash (BL) level for the components with relatively higher amount of retained (higher amounts of retained austenite led to the dimensional changes in the components during the service).

On the other hand, samples processed at optimized conditions (A2, B2, and C1) showed a coarse bainitic structure (**Figure 5(B)**), which is considered as one of the toughest structures of steel [22, 23]. Moreover, the surface of the sample was covered with tempered martensite, as shown in **Figure 5(C)**.

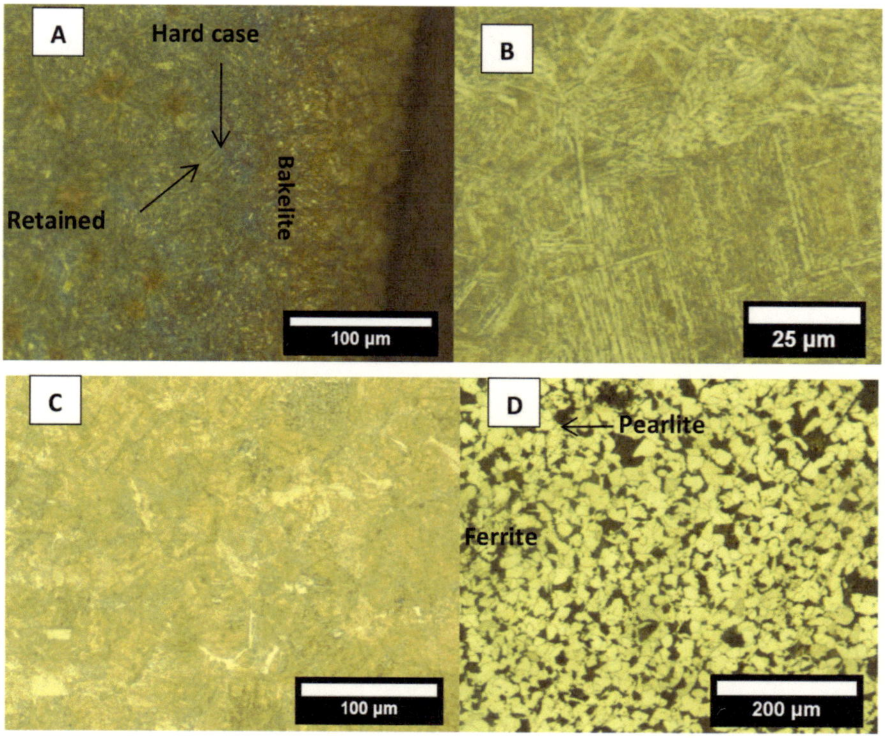

Figure 5. Optical microscope images of the samples: (A) microstructure of the sample processed at A1, B3, and C1 conditions; (B) microstructure at the core of the sample processed at A2, B2, and C1 conditions; (C) microstructure at the surface of the sample processed at A2, B2, and C1 conditions; and (D) microstructure of the SAE 8620 low-alloy steel prior to the carburizing.

Figure 5(D) shows the microstructure of the SAE 8620 alloy steel prior to the heat treatment. It was observed that the microstructure is comprised up of fine grains with relatively higher amount of pearlite dispersed with small amount of ferrite as well, as shown in **Figure 5(D)**. Thus, it can be concluded that the microstructure of the specimen is in the suitable range for the heat treatment process. In the present study, the grain size was measured by using ASTM standards whereas the amounts of ferrite (10%) and pearlite (90%) in the sample were analyzed by using the software equipped with LECO microscope.

3.3. Defect per million opportunities (DPMO) model

In order to gain deep insight of the optimized parameters predicted by the DoE array, DPMO was calculated by using the following formula [12]:

$$DPMO = \frac{\textit{no. of defects found in a sample}}{\textit{sample size} \times \textit{no. of defect opportunities per unit}} \times 10{,}00{,}000 \quad (2)$$

A number of defect opportunities were 24, which could be calculated by the help of cause and effect diagram (not reported here). In short, the cause and effect diagram takes into account both internal (environment, human error, machine error, machine constraints, etc.) and external factors (carburizing time, carburizing temperature, carbon potential, hardening time, etc.).

However, at least 10,000 experiments were conducted from the optimized set of parameters, that is, A2, B2, and C1, but no significant variations were observed. This may be due to the fact that DoE approach takes into account almost all factors at the same time. Thus, the chances for the deviations are minimized. DPMO for the optimized set of conditions was 83, which is quite good according to the Six Sigma approach. Moreover, considering the complexity of the gas carburizing process, it must be highlighted that achieving such a high reproducibility of the surface treatment process is a challenging task, which was accomplished by the DoE approach. The present study is believed to be helpful in further reducing the DPMO values upon further improvements in this study.

4. Conclusions

The conclusions of this chapter are listed as follows:

1. SAE 8620 low-alloy steel was successfully case hardened under various conditions.
2. Control parameters and their levels for the gas carburizing process were selected on the basis of fishbone diagram.
3. Taguchi design of experiment approach was applied to build L9 matrix.
4. Control parameters and their levels were studied in terms of ECD, core hardness, and surface hardness.
5. DoE approach elucidated that the best parameters are delay quenching interval of 60 seconds, hardening temperature of 800°C, and soaking time in oil for 180 seconds.
6. The samples processed at optimized parameters possessed coarse bainitic structure at core and tempered martensite at the surface.
7. The optimized parameters were highly reproducible, as evaluated by the DPMO model.

Author details

Muhammad Atiq Ur Rehman[1,4], Muhammad Azeem Munawar[2], Qaisar Nawaz[3] and Muhammad Yousaf Anwar[4]*

*Address all correspondence to: myanwar@uet.edu.pk

1 Institute of Space Technology Islamabad, Islamabad, Pakistan

2 Institute of Polymer Materials Friedrich-Alexander – University, Erlangen-Nuremberg, Erlangen, Germany

3 Department of Material Science and Engineering, Institute of Biomaterials, University of Erlangen Nuremberg, Erlangen, Germany

4 Department of Metallurgical and Materials Engineering, University of Engineering and Technology Lahore, Lahore, Pakistan

References

[1] Murugan VK, Mathews PK. Optimization of heat treatment processes using Taguchi's parameter design approach. International Journal of Research in Mechanical Engineering. 2013;**1**:16-21. ISSN Online: 2347-5188 Print: 2347-8772

[2] Tsui KL. An overview of Taguchi method and newly developed statistical methods for robust design. IIE Transaction on Institute Industrial Engineering. 1992;**24**:44-57. DOI: 10.1080/07408179208964244

[3] Rao RS, Kumar CG, Prakasham RS, Hobbs PJ. The Taguchi methodology as a statistical tool for biotechnological applications: A critical appraisal. Biotechnology Journal. 2008;**3**:510-523. DOI: 10.1002/biot.200700201

[4] Atiq Ur Rehman M, Bastan FE, Haider B, Boccaccini AR. Electrophoretic deposition of PEEK/bioactive glass composite coatings for orthopedic implants: A design of experiment (DoE) study. Materials and Design. 2017;**130**:223-230. DOI: 10.1016/j.matdes.2017.05.045

[5] Pishbin F, Simchi A, Ryan MP, Boccaccini AR. A study of the electrophoretic deposition of bioglass® suspensions using the Taguchi experimental design approach. Journal of the European Ceramic Society. 2010;**30**:2963-2970. DOI: 10.1016/j.jeurceramsoc.2010.03.004

[6] Corni I, Cannio M, Romagnoli M, Boccaccini AR. Application of a neural network approach to the electrophoretic deposition of PEEK–alumina composite coatings. Materials Research Bulletin. 2009;**44**:1494-1501. DOI: 10.1016/j.materresbull.2009.02.011

[7] Fatoba OS, Akanji OL, Aasa AS. Optimization of carburized UNS G10170 steel process parameters using Taguchi approach and response surface model (RSM). Journal of Minerals and Materials Characterization and Engineering. 2014;**02**:566-578

[8] Karabelchtchikova O. Fundamentals of mass transfer in gas carburizing [PhD thesis]. Worcester, USA: Worcester Polytechnic Institute; 2007

[9] Pishbin F, Simchi A, Ryan MP, Boccaccini AR. Electrophoretic deposition of chitosan/45S5 bioglass composite coatings for orthopaedic applications. Surface and Coatings Technology. 2011;**205**:5260-5268. DOI: 10.1016/j.surfcoat.2011.05.026

[10] Kumar S, Rakesh Kumar SA. Optimization of heat treatment processes of steel used in automotive bearings. International Journal of Technical Research & Applications. 2016;**4**:38-44

[11] Gerald JH, Doganaksoy N, Hoerl R. The evolution of six sigma. Quality Engineering Journal. 2000;**12**:317-326. doi.org/10.1080/08982110008962595

[12] Şenvar Ö, Tozan H. Process capability and six sigma methodology including fuzzy and lean approaches. In: Products And Services: From R&D To Final Solutions. Rijeka: Intech, Cop; 2010. pp. 153-178. Available from: http://www.intechopen.com/source/pdfs/12326/InTech-Process_capability_and_six_sigma_methodology_including_fuzzy_and_lean_approaches.pdf

[13] Boniardi M, D'Errico F, Tagliabue C. Influence of carburizing and nitriding on failure of gears – A case study. Engineering Failure Analysis. 2006;**13**:312-339. DOI: 10.1016/j.engfailanal.2005.02.021

[14] Bensely A, Stephen JS, Mohan LD, Nagarajan G, Rajadurai A. Failure investigation of crown wheel and pinion. Engineering Failure Analysis. 2006;**13**:1285-1292. DOI: 10.1016/j.engfailanal.2005.10.002

[15] Sugianto A, Narazaki M, Kogawara M, Shirayori A, Kim SY, Kubota S. Numerical simulation and experimental verification of carburizing-quenching process of SCr420H steel helical gear. Journal of Materials Processing Technology. 2009;**209**:3597-3609. DOI: 10.1016/j.jmatprotec.2008.08.017

[16] Asi O, Can AC, Pineault J, Belassel M. The relationship between case depth and bending fatigue strength of gas carburized SAE 8620 steel. Surface and Coatings Technology. 2007;**201**:5979-5987. DOI: 10.1016/j.surfcoat.2006.11.006

[17] Genel K. Effect of case depth on fatigue performance of AISI 8620 carburized steel. International Journal of Fatigue. 1999;**21**:207-212. DOI: 10.1016/S0142-1123(98)00061-9

[18] Asi O, Can AC, Pineault J, Belassel M. The effect of high temperature gas carburizing on bending fatigue strength of SAE 8620 steel. Materials & Design. 2009;**30**:1792-1797. DOI: 10.1016/j.matdes.2008.07.020

[19] Richman RH, Landgraf RW. Some effects of retained austenite on the fatigue resistance of carburized steel. Metallurgical Transactions A. 1975;**6**:955-964. DOI: 10.1007/BF02661347

[20] Jian-Min T, Yi-Zhong Z, Tian-Yi S, Hai-Jin D. The influence of retained austenite in high chromium cast iron on impact-abrasive wear. Wear. 1990;**135**:217-226. DOI: 10.1016/0043-1648(90)90026-7

[21] Palaniradja K, Alagumurthi N, Soundararajan V. Optimization of process variables in gas carburizing process: A Taguchi study with experimental investigation on SAE 8620 and AISI 3310 steels. Turkish Journal of Engineering and Environmental Sciences. 2005;**29**:279-284

[22] Avner SH. Introduction to Physical Metallurgy. 2nd ed. New York: McGraw Hill; 1974

[23] Callister W, Rethwisch D. Materials Science and Engineering: An Introduction. New York: Wiley; 2007. DOI: 10.1016/0025-5416(87)90343-0

[24] Zhang X, Tang J, Zhang X. An optimized hardness model for carburizing-quenching of low carbon alloy steel. Journal of Central South University. 2017;**24**:9-16. DOI: 10.1007/s11771-017-3403-2

[25] Karabelchtchikova O, Sisson RD. Carbon diffusion in steels: A numerical analysis based on direct integration of the flux. Journal of Phase Equilibria and Diffusion. 2006;**27**:598-604. DOI: 10.1007/BF02736561

[26] Singer F, Kufner M. Model based laser-ultrasound determination of hardness gradients of gas-carburized steel. NDT&E International. 2017;**88**:24-32. DOI: 10.1016/J.NDTEINT.2017.02.006

[27] Smirnov AE, Ryzhova MY, Semenov MY. Choice of boundary condition for solving the diffusion problem in simulation of the process of vacuum carburizing. Metal Science and Heat Treatment. 2017;**59**:237-242. DOI: 10.1007/s11041-017-0135-8

[28] Zhao J, Wang GX, Ye C, Dong Y. A numerical model coupling diffusion and grain growth in nanocrystalline materials. Computational Materials Science. 2017;**136**:243-252. DOI: 10.1016/j.commatsci.2017.05.010

[29] Tao Q, Wang J, Fu L, Chen Z, Shen C, Zhang D. Ultrahigh hardness of carbon steel surface realized by novel solid carburizing with rapid diffusion of carbon nanostructures. Journal of Materials Science and Technology. 2017;**33**:1210-1218. DOI: 10.1016/J.JMST.2017.04.022

[30] Easton D, Perez M. Effects of forming route and heat treatment on the distortion behaviour of case-hardened martensitic steel type S156. In: Heat Treatment. Ohio: At Columbus; 2017

[31] Peng YW, Gong JM, Jiang Y, Fu MH, Rong DS. Influence of plastic pre-strain on low-temperature gas carburization of 316L austenitic stainless steel. Applied Mechanics and Materials. 2016;**853**:178-183. DOI: 10.4028/www.scientific.net/AMM.853.178

[32] Dal'Maz Silva W, Dulcy J, Ghanbaja J, Redjaïmia A, Michel G, Thibault S, Belmonte T. Carbonitriding of low alloy steels: Mechanical and metallurgical responses. Materials Science and Engineering A. 2017;**693**:225-232. DOI: 10.1016/j.msea.2017.03.077

[33] Liew WYH, Ling JLJ, Siambun NJ. Sliding wear behaviour of steel carburized using Na_2CO_3–NaCl. MATEC Web Conferences. 2017;**87**:2010. DOI: 10.1051/matecconf/20178702010

[34] Palaniradja K, Alagumurthi N, Soundararajan V. Hardness and case depth analysis through optimization techniques in surface hardening processes. The Open Materials Science Journal. 2010;**4**:38-63

Chapter 8

Development of Falling Film Heat Transfer Coefficient for Industrial Chemical Processes Evaporator Design

Muhammad Wakil Shahzad,
Muhammad Burhan and Kim Choon Ng

Additional information is available at the end of the chapter

http://dx.doi.org/10.5772/intechopen.69299

Abstract

In falling film evaporators, the overall heat transfer coefficient is controlled by film thickness, velocity, liquid properties and the temperature differential across the film layer. This chapter presents the heat transfer behaviour for evaporative film boiling on horizontal tubes, but working at low pressures of 0.93–3.60 kPa as well as seawater salinity of 15,000–90,000 mg/l or ppm. Owing to a dearth of literature on film-boiling at these conditions, the chapter is motivated by the importance of evaporative film-boiling in the process industries. It is observed that in addition to the above-mentioned parameters, evaporative heat transfer of seawater is affected by the emergence of micro-bubbles within the thin film layer, particularly when the liquid saturation temperatures drop below 25°C (3.1 kPa). Such micro-bubbles are generated near to the tube wall surfaces, and they enhanced the heat transfer by two or more folds when compared with the predictions of conventional evaporative film-boiling. The appearance of micro-bubbles is attributed to the rapid increase in the specific volume of vapour, i.e. dv/dT, at low saturation temperature conditions. A new correlation is thus proposed in this chapter and it shows good agreement to the measured data with an experimental uncertainty less than ±8%.

Keywords: low pressure evaporation, falling film evaporation, horizontal tubes evaporators

1. Background

In process industries such as the refineries, food and desalination plants, the need of high-performance evaporators is paramount to minimize irreversibilities due to high heat transfer as well as to reduce footprint area of associated components. A falling film evaporator is one of the key design components which are associated with not only high heat transfer rates but are also immune to change in feed qualities. In particular, for present desalination application, the falling film evaporative process could augment heat transfer rates involving brines which inherently reduce the equipment cost because of compact design.

In this chapter, a horizontal tube falling film evaporator is studied for low-temperature applications, particularly for the desalination industry. The first part of this chapter focuses on advantages of horizontal tube falling film evaporators over flooded evaporators and vertical tube evaporators and its applications. In the second part of the chapter, a literature review on falling film evaporation heat transfer coefficient (FFHTC) to the extent necessary for this work is provided. A novel FFHTC for low-temperature (below ambient) applications and for different salt concentrations is developed in the third part of this chapter. The comparison of proposed correlation with traditional Han and Fletcher [1] correlation and the effect of different operational parameters on heat transfer is discussed in the last section of the chapter.

Flooded evaporators have been used in desalination industry for long time. Recently, there is a thrust of horizontal tubes falling film evaporators over the flooded evaporators because of their advantages. They also replaced the vertical tube evaporators because of its unique characteristics. Falling film evaporators in general, are highly responsive to operational parameters, such as energy supply, pressure levels, feed rate, and salt concentrations in the feed. The fact that falling film evaporators can be operated across small temperature differences make them amenable to the application in multiple effect configurations. The advantages of falling film evaporators are outlined in the section below.

2. Advantages of falling film evaporators

The main advantages of falling film evaporators over flooded evaporators are as follows:

1. High heat transfer coefficient and resulting compact design.
2. More uniform overall heat transfer coefficient value across the tube bundle.
3. Reduction in working fluid requirement to about one-third as compared to flooded evaporators.
4. Short product contact times, typically just a few seconds per pass.
5. Minimization of salt deposition on tubes surface that helps in cleaning the tubes.

The potential advantages of horizontal tube evaporators over vertical tubes evaporators are as follows:

1. Heat transfer coefficients for horizontal tubes are higher than those for vertical tubes since the heated flow length is much shorter.
2. External enhancements are available for tubes in copper, copper-nickel and stainless steel, etc. for up to a 10-fold increase in evaporation coefficient.

3. A horizontal tube bundle can have multiple tube passes of the heating fluid to significantly increase its heat transfer coefficient as compared to vertical tubes evaporators with single pass.
4. A larger Length to diameter ratio (L/D) ratio horizontal shell evaporator can be designed as compared to small L/D ratio of vertical evaporator that helps to prevent the dry out and flooding in the tubes.
5. The two pass (U-tube) design in horizontal tube evaporators is much more efficient, cheaper and easier to maintain compared to the single pass floating head in vertical tubes.
6. Flow length of liquid film in a horizontal tube evaporator minimizes the liquid hold-up time and residence time during operation.
7. Horizontal tubes bundle arrangement reduces the unit height that helps to reduce the piping work.
8. Horizontal arrangement reduces the footprint for large-capacity plant because the evaporators can be arranged in double tier arrangement.

Although the horizontal tubes falling film evaporators have advantages over flooded and vertical tubes evaporators, the main limitation is the lack of heat transfer data particularly at low temperature, i.e., below 323 K.

3. Heat transfer review for falling film evaporators

A critical appreciation of the thermal performance is essential for the optimum design of falling film horizontal tube evaporators especially for desalination industry. A large number of empirical and theoretical heat transfer coefficients correlations are available in literature. The majority of those available correlations are for different refrigerants, and few of them are for pure water and limited to saturation temperatures more than 323 K.

Many researchers provided the detailed overview of available correlations. A critical review is published by Ribatski and Jacobi [2] who tabulated the heat transfer correlations in terms of dimensionless numbers as developed by many researchers. They also provided heat transfer coefficient values for water and different refrigerants with single-tube and multi-tube evaporators. They concluded that every correlation has a limited validation governed by operating parameters under which they developed, and efforts are needed to generalize these correlations. Adib et al. [3] conducted the experiment with vertical tube falling film evaporator, and they calculated the heat transfer coefficient value using correlation available in literature [4–8] and found good agreement with experimental results. Uche et al. [9] compared the heat transfer correlations at different inlet brine temperatures and for different mass velocities for horizontal and vertical tube evaporators. They also compared their results with different available correlations [1, 10–14] and found that Parken correlation can be used for nonboiling conditions, and Han and Fletcher's correlation is good for boiling conditions. A falling film evaporation analytical model is developed by Fujita et al. [15–17] using R-11, and they analysed the drip pattern, droplets, and sheet modes. They found that accuracy of their model is within ±20%.

Table 1 summaries heat transfer correlations of many researchers found in the literature. This table also highlights the limitations of applications of these correlations, such as the types of working fluids, pressures and the temperature ranges and evaporator geometry.

Since operational and design parameters are the key factors to maximize the evaporator performance, so researches are provided extensive data on it. Film modes are controlled by film Reynolds numbers, and different heat transfer coefficient behaviours are noticed by researchers for smooth tubes as Reynolds number changes [15, 18–21]. They observed three kinds of behaviour such as (1) heat transfer coefficient decreases to its minimum value and then increases again, (2) it increases with Reynolds number, and (3) heat transfer coefficient increases to its maximum value and then drops. Lorenz and Yung [22] investigated that film evaporation on a single tube is different to an array of tubes, and it may be due to turbulence of inter-tube evaporation. They also found that critical Reynolds number affects the evaporation heat transfer, and for below 300, the heat transfer coefficient value for a single tube is higher as compared to an array of tubes. Thome et al. [23] conducted the experiments for falling film heat transfer coefficient for four types of tubes, such as plain, turbo-BII HP, Gewa-B and high-flux tubes. They concluded that for different inter-tube flow modes, there is no discernible difference in heat transfer coefficients in respective flow zone. Fujita et al. [15] investigated that the heat transfer value is low on the top row of tubes which is due to direct expose to feed supply. They also investigated the effect of feeder type on heat transfer coefficient. They used refrigerant R-11 on horizontal tube evaporators. Liu et al. [18] performed falling film heat transfer experiments for different tubes surfaces, and they concluded that the value is from 3- to 4-folds higher for roll-worked tubes as compared to smooth tubes. They also found that both the flow conditions and tubes spacing have negligible effect on the heat transfer coefficient. Aly et al. [24] conducted the tests for deposit film thickness effects, and they found drastic decrease in heat transfer with increase in deposition thickness. Moeykens et al. [25, 26] and Chang et al. [27] performed falling film experiment tests for R-123, R-134a, R-22, and R-141b, and they found that it can be enhanced by adding the collection tray under each tube row. The falling film correlations developed by researchers [26, 28–30] for refrigerants R-22, R-123, R-134a, and R-141b are having uncertainty of 20–25% by using four different apparatuses. Bourouni et al. [31] performed the experiments with aero-evaporator, and they reported that increase in characteristic dimensions of heat exchanger results in a significant increase in the evaporative performance. Yang and Shen [32] found that the heat transfer coefficient is a strong function of heat input and increases with heat input. The vapour flow effect due to liquid drag and dry out of tubes is studied by Ribatski and Jacobi [2]. The effect of dynamics of film on heat transfer is investigated by Xu et al. [33] and Yang and Shen [34]. They found that increase in liquid load causes perturbation in film that enhances the heat transfer. They also reported that increase in tube diameter does not favour heat transfer which can be due to more turbulence in film on smaller diameter tubes. For horizontal tubes falling film evaporators, Han and Fletcher [1] is the most famous correlation, whereas Chun and Seban [35] is used for vertical tube. Both of these famous correlations are for pure water and for saturation temperatures of 322 K or more.

References	Correlation
Xu et al. [33]	$h_{\text{evaporation}} = 05.169 \times 10^{-11} \left[\frac{h_{fg} \cdot g \cdot \rho_1^2 \cdot D^2}{\Delta t^2 \cdot \mu_l} \right]^{-0.333} \left(\frac{\overline{\delta}}{D} \right)^{-0.422\Delta t^{0.503}} \left(1 + \frac{\delta_{\max} - \delta_{\min}}{\overline{\delta}} \right)$ Deionized water, 50°C, horizontal copper tubes evaporator.
Fujita et al. [15]	for 1st tube: $Nu = \left((\text{Re}_f)^{-1/3} + 0.008(\text{Re}_f)^{0.3}(\text{Pr})^{0.25} \right)^{1/2}$ For 2nd to 5th tubes: $Nu = \left((\text{Re}_f)^{-1/3} + 0.01(\text{Re}_f)^{0.3}(\text{Pr})^{0.25} \right)^{1/2}$ Freon R-11, electrically heated five horizontal copper tubes, OD-25 mm
Han and Fletcher [1]	$h_{\text{evaporation}} = 0.0028. \left[\frac{\mu_l^2}{g.\rho_l^2.k_l^3} \right]^{-0.333} (\text{Re}_\Gamma)^{0.5}(\text{Pr})^{0.85}$ Pure water, 49–127°C, electrically heated single horizontal tube, OD-50.8 mm, thickness-1.7 mm, Length-254 mm
Bourouni et al. [31]	$h_f = 2.2. \left[\frac{v_f^2}{g.k_l^3} \right]^{-0.333} . \left(\frac{H}{OD} \right)^{0.1} .(\text{Re}_f)^{-0.333}$ Pure water, 60 and 90°C, Polypropylene horizontal tubes aero-evaporator , OD-25.4 mm,

References	Correlation
Chun and Seban [35]	$h_{\text{film}} = 0.821.\left[\frac{\mu_l^2}{g.\rho_l^2.k_l^3}\right]^{-0.333} (\text{Re}_\Gamma)^{-0.22}$
	Pure water, 46–118°C, electrically heated single vertical tube, OD-28.58 mm, thickness-0.1 mm, Length-292 mm
Alhusseini et al. [36]	For laminar region:
	$h^*{}_{\text{laminar}} = 2.65.(\text{Re})^{-0.158}(Ka)^{0.0563}\lim_{x\to\infty}$
	For combine:
	$h^* = (h^5{}_{\text{laminar}} + h^5{}_{\text{turbulent}})^{1/5}$
	Propylene glycol and water
Shmerler et al. [37]	$h_E^* = 0.0038.(\text{Re}_\Gamma)^{0.35}(\text{Pr})^{0.95}$
	Pure water, electrically heated single vertical tube, OD-25.40 mm, Length-781 mm
Chien et al. [38]	$Nu_{cv} = 0.0386.(\text{Re}_\Gamma)^{0.09}.(\text{Re}_\Gamma)^{0.986}$
	R245fa, 5 and 20°C, horizontal smooth tubes.

Table 1. Review of heat transfer coefficient correlations for different evaporator design and operation conditions.

It can be seen from the above discussion that Han and Fletcher's correlation is most frequently used for film boiling on horizontal tubes. This correlation is developed with pure water evaporating at temperatures 322 K and above. There is a lack of data for evaporative film boiling typically below ambient condition. The boiling data pertaining to saline solution of 15,000–90,000 mg/l or ppm are also scarce, and yet these conditions are particularly important for the designing of falling film evaporators for processes industries and desalination plants, such as food and beverage, multi-effect desalination (MED) and multi-stage flash evaporation (MSF). Many manufacturers, perhaps due to competition reason, are not revealing their proprietary film boiling data at these conditions. We designed experiments to develop falling film heat transfer coefficient for low-temperature evaporator typically from 279 to 300 K and pressure from 0.93 to 3.60 kPa. The new proposed correlation will be applicable for wide range of concentration evaporator design. We also presented the effect of salt concentration on heat transfer and log mean temperature difference (LMTD). The proposed designed experiments will help process industries to design falling film evaporators for wide range of operation.

4. Falling film heat transfer coefficient development

The methodology used here is to adopt Han and Fletcher's correlation for film boiling on horizontal tubes and to enhance its use by incorporating the effects of salinity and by expanding the range of temperatures of its application for horizontal tubes falling film evaporation.

4.1. Theoretical model

The non-dimensional terms in Han and Fletcher correlation model, namely, the Reynolds, Prandtl and Nusselt numbers are adequate to describe the surface evaporation from liquid film due to thermal effect. At low saturation pressures, the vapour specific volume rapidly increases, and this could possibly leads to enhancement of heat transfer. Han and Fletcher model is revisited to capture this additional heat transfer enhancement phenomenon. At a low saturation temperature the micro-bubble generated at tube surface can lift up quickly because of high specific volume and break through the thermal barrier within liquid film. The traditional heat transfer models are unable to define this augmentation of heat transfer enhancement by buoyancy fortified bubble agitation.

The Han and Fletcher correlation given in **Table 1** can also be expressed in a more familiar form as shown in Eq. 1

$$\frac{h_{\mathrm{evap}}\left(\frac{\mu_l^2}{g\rho_l^2}\right)^{1/3}}{k_l} = Nu = 0.0028(\mathrm{Re}_\Gamma)^{0.5}(\mathrm{Pr})^{0.85} \tag{1}$$

where indices and the constant term are found for the boundary conditions of film boiling. For the determination of the overall heat transfer coefficient, the total heat transfer is computed via heat transferred to circulating water, i.e.

$$Q_{in} = \dot{m}_{ch,w}\ Cp_{ch,w}\ (T^{o}_{ch,w} - T^{i}_{ch,w}) \tag{2}$$

Using the concept of log mean temperature difference (LMTD) and the saturation temperature of evaporator, the overall heat transfer coefficient ($U_{overall}$) of the evaporator can be expressed as

$$UA_{overall} = \frac{\dot{m}_{ch,w}\ Cp_{ch,w}\ (T^{out}_{ch,w} - T^{in}_{ch,w})}{\left\{\frac{(T^{out}_{ch,w} - T_{sat}) - (T^{in}_{ch,w} - T_{sat})}{\ln \frac{(T^{out}_{ch,w} - T_{sat})}{(T^{in}_{ch,w} - T_{sat})}}\right\}} \tag{3}$$

The local falling film heat transfer coefficient on film side (h) is deduced from the knowledge of the resistance due to chilled water flow inside the tubes which is calculated by the Dittus-Boelter correlation given in Eq. 4

$$Nu = 0.023\mathrm{Re}^{0.25}\mathrm{Pr}^{n} \tag{4}$$

The pipe wall resistance (stainless steel 316) is negligible due to small thickness (0.7 mm). The evaporation heat transfer coefficient is calculated by using overall heat transfer coefficient given in Eq. 5

$$\frac{1}{UA} = \left(\frac{1}{hA}\right)_{tubeside} + R_{wall} + \left(\frac{1}{hA}\right)_{outside} \tag{5}$$

The experimental program is planned for capturing the two unknown parameters in above Eq. 5.

4.2. Experimental apparatus

Adsorption desalination (AD) plant existing in air-conditioning laboratory is used to conduct the experiments. **Figures 1** and **2** show the AD plant installed in National University of Singapore (NUS) and plant operational schematic.

There are five main components of AD plant namely: (1) evaporator, (2) adsorber/desorber beds, (3) condenser, (4) conditioning facility and (5) pre-treatment facility. The evaporator shell and tubes are fabricated with stainless steel and are arranged horizontally details of which are shown in **Figure 3**.

The evaporator tubes are arranged in four rows with 12 tubes in each row. This evaporator is 4 pass using a 'water box' arrangement at the ends of the heat exchanger. Special profiled tubes are used in evaporator to enhance the heat transfer. The details of the tube are shown in **Figure 4**.

A precise electrical thyristor controller is installed to supply the chilled water to evaporator at constant inlet temperature. This thyristor maintains the temperature fluctuations at inlet of coolant water to less than ±0.15 K. The chilled water supply is regulated at 48 l/min. Since experiments are conducted at different salt concentrations, and constant salt concentration

http://dx.doi.org/10.5772/intechopen.69299

Figure 1. Pictorial view of adsorption desalination plant installed in NUS.

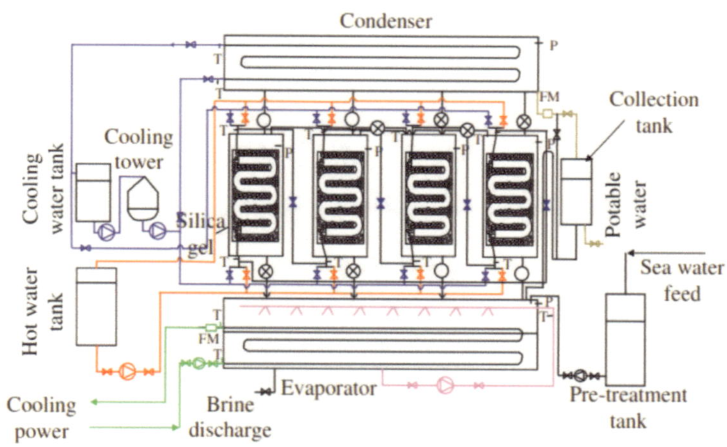

Figure 2. Detailed operational schematic of adsorption desalination plant schematic.

condition in evaporator is maintained by re-circulation of the condensate back to evaporator via U-tube. To maintain a constant liquid film on tube surface, a spray pump is used to discharge fine water droplets (nominally 0.1–0.15 mm diameter) through nozzles on top of tube bundle. The design parameters of evaporator are given in **Table 2**.

4.2.1. Experimental procedure

Experimental procedure can be categorized into operation of individual components namely: (1) evaporator, (2) Vacuum system, (3) adsorber/desorber and (4) condenser.

The evaporator operation can be divided into two circuits namely: (1) feed water circuit and (2) chilled water circuit.

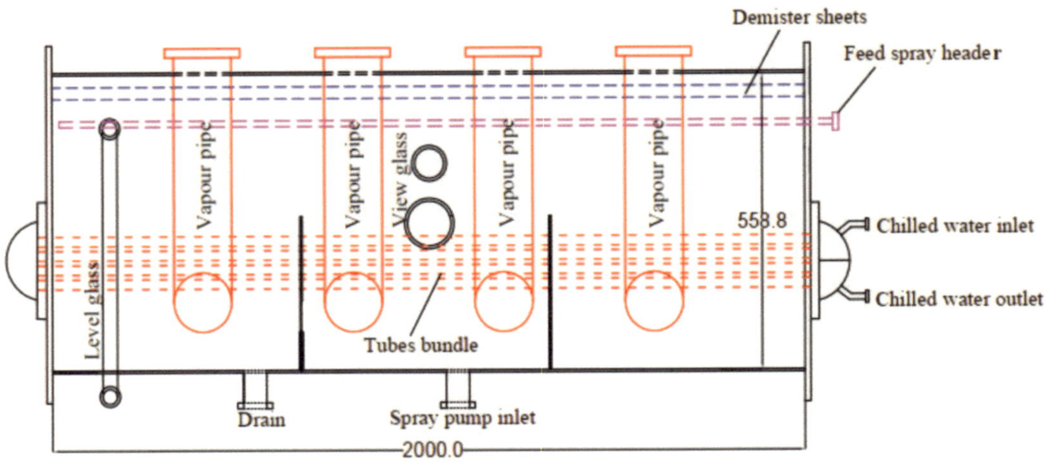

Figure 3. Adsorption desalination cycle evaporator detailed design.

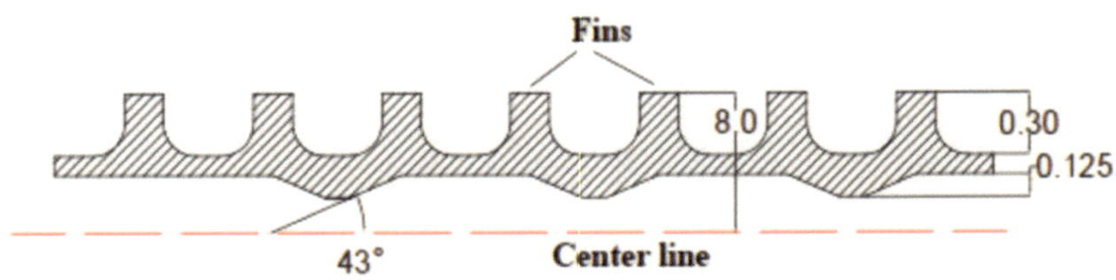

Figure 4. Cross section of end-cross tube used in evaporator of adsorption desalination plant.

Parameters	Values	Units
Number of tubes	48	
Length of each tube	1900	mm
Tube outer diameter	16	mm
Tube thickness	0.7	mm
No of passes	4	
Shell diameter	558.8	mm
Shell length	2000	mm

Table 2. Design parameters of adsorption desalination system evaporator.

4.2.1.1. Feed water circuit

The seawater/feed first enters into a pre-treatment facility to remove particulates and suspensions and then to the de-aeration tank to de-aerate. In the de-aeration tank, the dissolved non-condensable is removed before the feed enters to AD evaporator. The de-aerated feed is then pumped into the evaporator via feed pump. A spray pump is installed with evaporator to spray the feed on to the tube bundle via spray nozzles. This is special magnetic pump that can operate in vacuum environment. The reflux from condenser maintains the salt concentration level inside the evaporator. This feed water line is provided with flow meter and valve to regulate the feed flow.

4.2.1.2. *Chilled water circuit*

The chilled water is the heat source that is circulated inside the tubes of evaporator. An electrical heater is installed to maintain the coolant temperature. This heater is controlled by a thyristor controller to maintain its inlet temperature. Chilled water circuit is equipped with regulating valve and flow meter to adjust the flow rate such that the evaporator can be operated under different conditions. The operation parameters are given in **Table 3**.

4.2.1.3. *Vacuum system*

A water vapour tolerant vacuum pump is necessary since the operation of AD system is under vacuum. Prior running an experiment vacuum holding capacity of the system is tested for 36 h, and it is found that the vacuum leak is negligible. During an experiment vacuum pump helps to maintain the desired saturation pressure inside the evaporator by pulling the air in case it ingress into the system. To ensure that the film on the tube surface is evaporating all the time, it is imminent to maintain the saturation temperature which is always lower than chilled water temperature inside the tubes.

4.2.1.4. *Adsorber/desorber bed operation*

The evaporator is connected to adsorber bed filled with silica gel via pneumatic valves to adsorb the water vapour. The adsorption of water vapour sustains the continuous evaporation in the evaporator. The heat of adsorption is removed by circulation of cooling water inside the adsorber coolant flow channel.

Similarly, a desorber bed is connected to a condenser and heat of desorption is supplied by a heater controlled by a thyristor controller.

4.2.1.5. *Condenser operation*

The desorber bed is connected to a condenser where the desorbed vapours are condensed on shell side. The cooling water circulated through the tubes of condenser is regenerated in a cooling tower at roof top.

Parameters	**Values**	**Units**
Chilled water flow rate	48	LPM
Sea water flow rate (Γ)	1.1	LPM/m of tube length
Evaporator saturation temperature	279–300	K
Evaporator saturation pressure	0.93–3.60	kPa
Feed water salinity range	15,000–90,000	ppm

Table 3. Operational parameters of adsorption desalination cycle.

The apparatus is fully instrumented to capture all required data. A Yokogawa pressure transmitter of range 0–60 KPa abs. (accuracy ±0.25%) is installed on the evaporator for saturation pressure readings. The OMEGA 5 kΩ type thermistors (accuracy ±0.15 K) are used for all temperature measurements. The KROHNE Flow meters (accuracy ±0.5% of reading) are used for flow measurements. All temperature, pressure and flow readings are continuously monitored by a data logger unit at intervals of 1 min.

A high speed camera is installed on the evaporator to observe the film behaviour over the tubes. It is observed that there is ample turbulence in liquid film on the tubes due to bubble formation on tube surface. The evidence of film turbulence is captured by camera shown in **Figure 5**, and more clear explanation by a film model is also presented.

There is a natural temperature gradient within liquid film on the tubes and the micro-bubble generation on tube surface agitates the liquid film when it tries to break through the thermal barrier. The micro-bubble generation and agitation phenomenon is explained in **Figure 6**. This

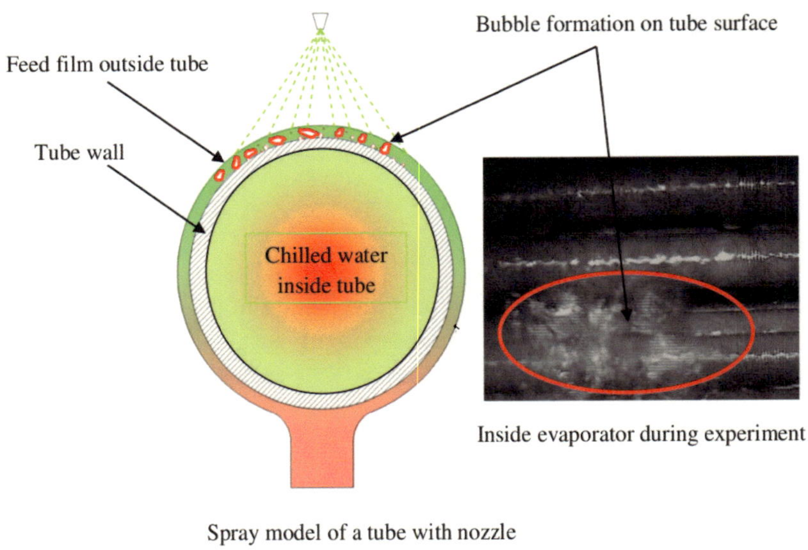

Figure 5. Bubbles formation in liquid film on tube surfaces and film agitation effect captured by camera.

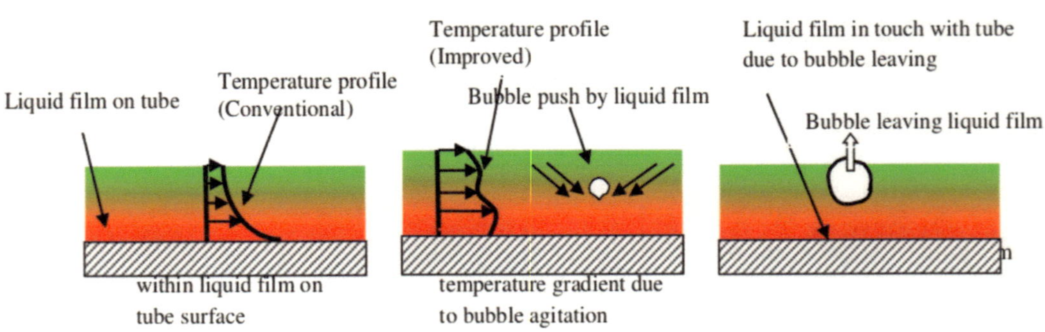

Figure 6. Film agitations due to bubbles movement and effect on conventional thermal gradient.

bubble agitation has two useful effects: first, it breaks the thermal barrier between the liquid film and tube surface that enhances the local heat transfer coefficient and second, when a micro-bubble moves up to the tube surface due to its very high specific volume, it also draw the heat from tube surface which further helps to enhance the heat transfer. An additional benefit is agitation within the liquid film due to the bubble movement.

5. Results and discussion

Figure 7 shows the experimental overall heat transfer coefficient values. The heat source temperatures vary from 10 to 40°C and salt concentration is 45,000 ppm. It can be seen from the results that overall heat transfer first drop with increase in chilled water temperature and then increase again at 40°C. A similar overall heat transfer trend is observed for 60,000 ppm (60 ppt) salt concentration as shown in **Figure 8**.

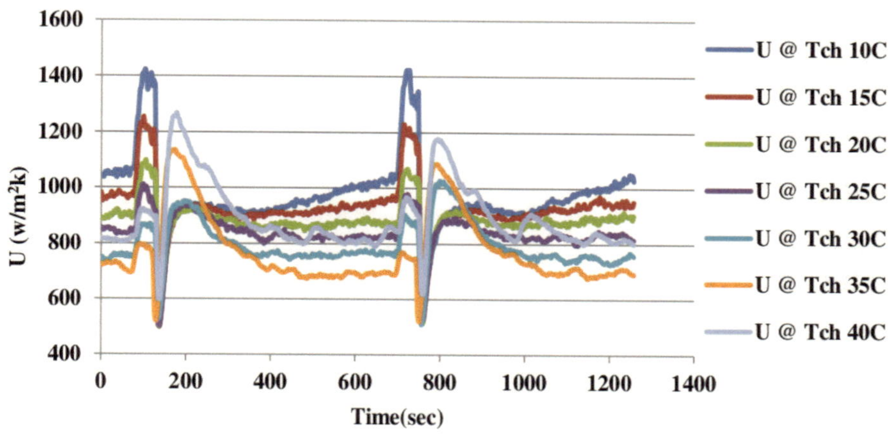

Figure 7. Typical experimental overall heat transfer coefficient profiles at 45000 ppm salt concentration.

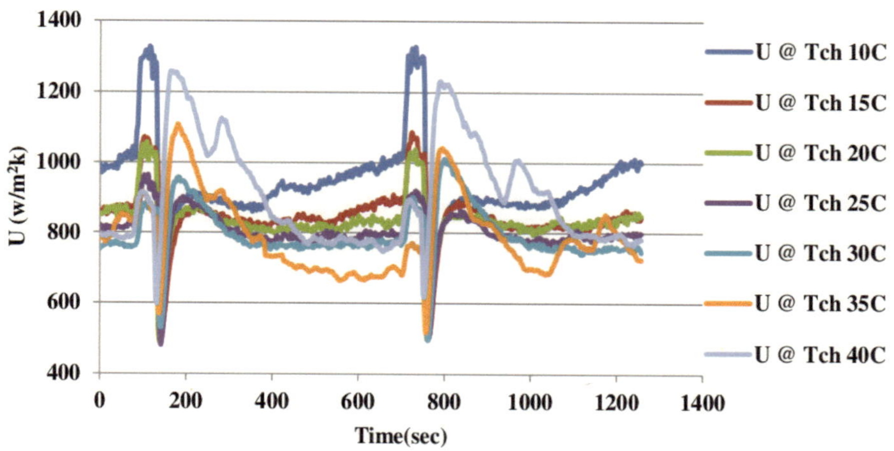

Figure 8. Typical experimental overall heat transfer coefficient profiles at 60000 ppm salt concentration.

The saturation temperature of evaporator and overall heat transfer coefficient values from experimental data at different chilled water inlet temperature and at different salt concentration are tabulated as shown in **Table 4**.

The evaporative heat transfer coefficient is calculated from experimental overall heat transfer coefficient by formulation as explained in theoretical model section. **Figure 9** shows the three-dimensional plot of evaporative heat transfer coefficients for assorted evaporator saturation temperatures and salinity level.

It can be seen from the plot that the heat transfer coefficient varies with saturation temperature and with salt concentration. It can be observed that at any salt concentration, it approaches the minimum value at 295 K and then with further decrease in saturation temperature the evaporation heat transfer coefficient value increases very sharply. It is also observed that specific volume of vapour increases very rapidly below at 295 K and above that temperature the change in specific volume of vapour is very small as shown in **Figure 10**.

It can be concluded that the sharp increase in evaporator heat transfer coefficient below 295 K may be due to bubble agitation. The micro-bubble produced on the tube surface from within the liquid film moves up quickly due to its very high specific volume and breaks the thermal barrier due to film agitation. This unique phenomenon is called 'bubble assisted evaporation'.

In film evaporation, 'micro-bubble agitation' plays an important role to enhance the heat transfer by reducing the thermal resistance between the liquid and tube surface barrier (model is shown in **Figure 6**). The traditional falling film evaporation heat transfer coefficient

Salinity	$T_{ch,in}$	T_{evap}	U	**Salinity**	$T_{ch,in}$	T_{evap}	U
	C	C	W/m²K		C	C	W/m²K
15,000	**10**	5.9	1025.45	**60,000**	**10**	5.9	937.61
	20	13.1	953.28		**20**	13.3	833.69
	30	20.3	885.17		**30**	19.7	776.62
	40	27.3	963.33		**40**	26.2	896.47
30,000	**10**	5.9	998.31	**75,000**	**10**	5.9	848.06
	20	13.1	920.78		**20**	13.0	751.47
	30	19.7	853.40		**30**	19.6	733.78
	40	25.7	906.96		**40**	26.9	893.53
45,000	**10**	5.6	970.78	**90,000**	**10**	5.5	815.94
	20	12.9	881.81		**20**	12.9	728.17
	30	19.3	798.17		**30**	19.3	694.79
	40	25.1	895.15		**40**	27.3	898.97

Table 4. Experimental overall heat transfer coefficient values and different saturation temperatures and at different salt concentrations.

http://dx.doi.org/10.5772/intechopen.69299

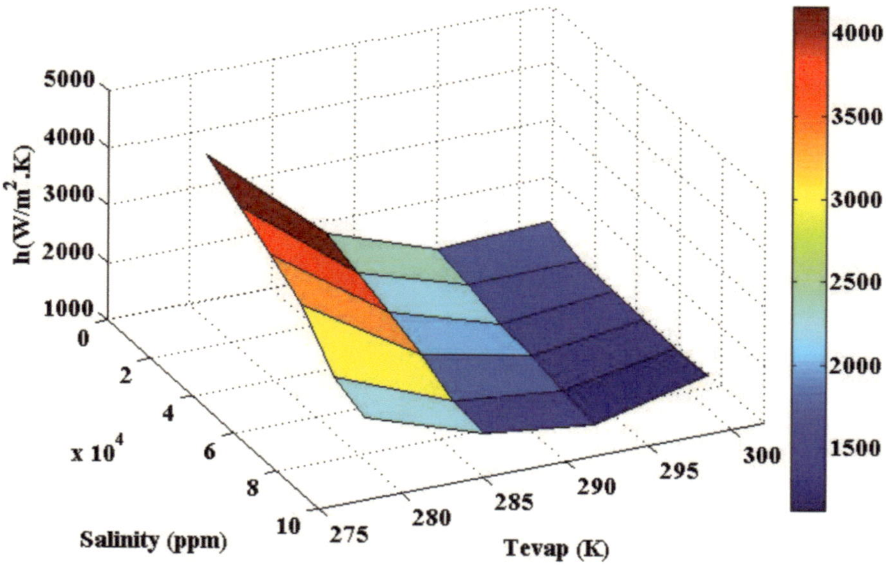

Figure 9. Experimental film evaporation heat transfer coefficient profiles at different saturation temperature and different salt concentrations.

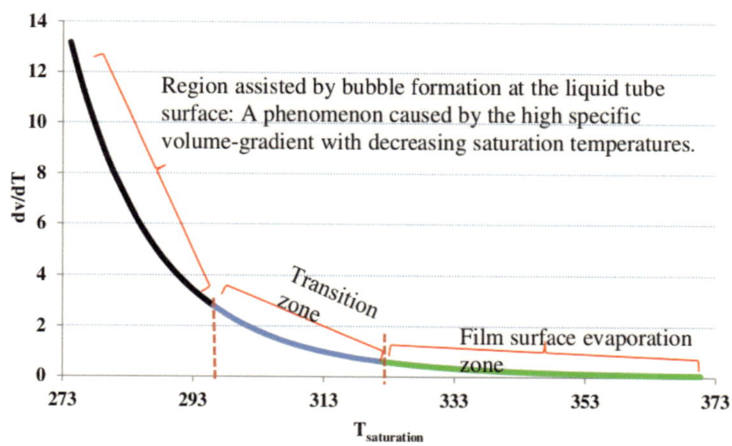

Figure 10. Change in vapour-specific volume with saturation temperature.

correlations (i.e. Han and Fletcher) do not capture this unique phenomenon and only capture the thermal driven film evaporation at saturation temperatures greater than 322 K.

A new falling film heat transfer coefficient with inclusion of 'bubble-assisted evaporation' for application at low saturation temperatures is proposed based on the experimental data. The above presented models (Eqs. 1–5) were written in FORTRAN to develop new correlation. The operational parameters namely: film velocity, salt concentration and heat flux are also included as additional parameters in the new correlation. In addition, to capture the effect of vapour specific volume, the gas volume term is also incorporated. The new correlation is given in

Eq. 3.6. **Figure 11** shows a comparison of Eq. 6 against the experimental data. It can be seen that new correlation has good agreement with experimental result. The measured heat transfer coefficient from experimental data has uncertainty of less than 8%. The Root mean square (RMS) error of regressed data is 3.5%. The additional terms used in the proposed correlation permit the limits of salinity and temperature to be accounted for, and a reference temperature, T_{ref} is taken as the reference temperature to match the region of Han and Fletcher.

$$h_{evap} = \left[0.277\left[\frac{\mu_l^2}{g.\rho_l^2.k_l^3}\right]^{-0.333}(Re_\Gamma)^{-2.11}(Pr)^{4.55}\left[2.\exp\left(\frac{S}{S_{ref}}\right)-1\right]^{-0.41}\left(\frac{T_{sat}}{T_{ref}}\right)^{14.70}\right]+ \left[0.885.\left(\frac{q}{\Delta T}\right)^{1}.\left(\frac{v_g}{v_{ref}}\right)^{-0.34}\right] \tag{6}$$

The above correlation is suitable for sub-atmospheric conditions from 0.93 to 3.60 kPa (corresponding to saturation temperatures 279–300 K) and feed water salinity ranges from 15,000 to 90,000 ppm. The film Reynolds number ranges $45< Re\Gamma < 90$ and Prandtl number ranges $5< Pr < 10$. In proposed superposition of effects in correlation, the first term is for film surface evaporation thermally driven and the second term is due to enhancement by the bubble assisted boiling effect.

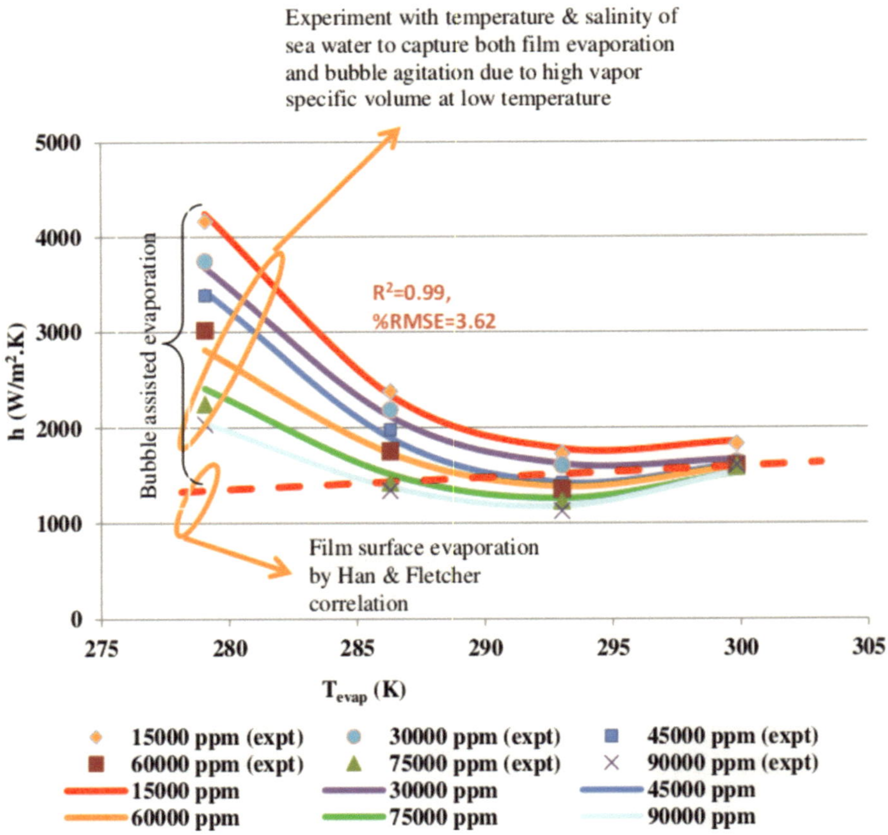

Figure 11. Falling film heat transfer coefficients values: experimental and proposed correlation.

The proposed falling film heat transfer coefficient is compared with Han and Fletcher correlation extrapolated to a region outside its validation range. The Han and Fletcher correlation is for pure water. It can be seen from **Figure 12** that Han and Fletcher correlation is only suitable for thermally driven surface evaporation for saturation temperatures 322 K and above.

A unique feature of the present correlation is the capture of 'bubble-assisted evaporation' which boosts the heat transfer coefficient by two to three folds at low saturation temperature. This additional effect seems to be significant only at a low saturation temperature 295 K or below. As a consequence, for situations where cooling and desalination are required simultaneously, the design of such an evaporator is likely to be more compact than at present.

This proposed falling film heat transfer coefficient is useful for falling film evaporator design for the process industries. It also includes concentration factor to accommodate operational variables for proper heat transfer area design.

The effects of operational parameters namely: (1) salt concentration and (2) saturation temperature on heat input and LMTD are also investigated. **Figure 13** shows the effect of these parameters on heat input. It can be seen that heat input increases with saturation temperature and it is due to increase in temperature difference of heat source. It can also be observed that salt concentration effect is negligible on heat input. **Figure 14** shows the effect of saturation temperature and salt concentration on LMTD. It can be observed that LMTD also increases with saturation temperature which is due to higher temperature differences at high saturation temperatures.

The salt concentration effect is minimal as can be seen from plot. The measured accuracy of log mean temperature difference (LMTD) and the heat input (Q) is 8%.

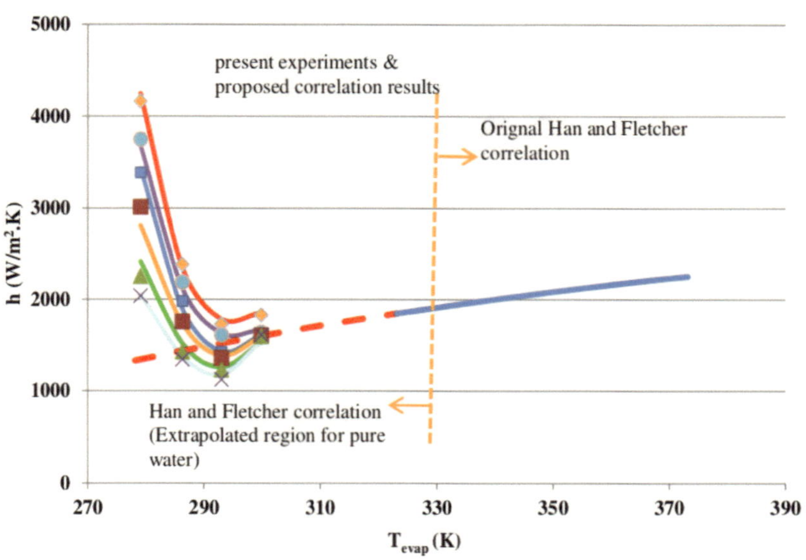

Figure 12. Falling film heat transfer coefficient values: experimental and proposed correlation compared with Han and Fletcher correlation extrapolated region.

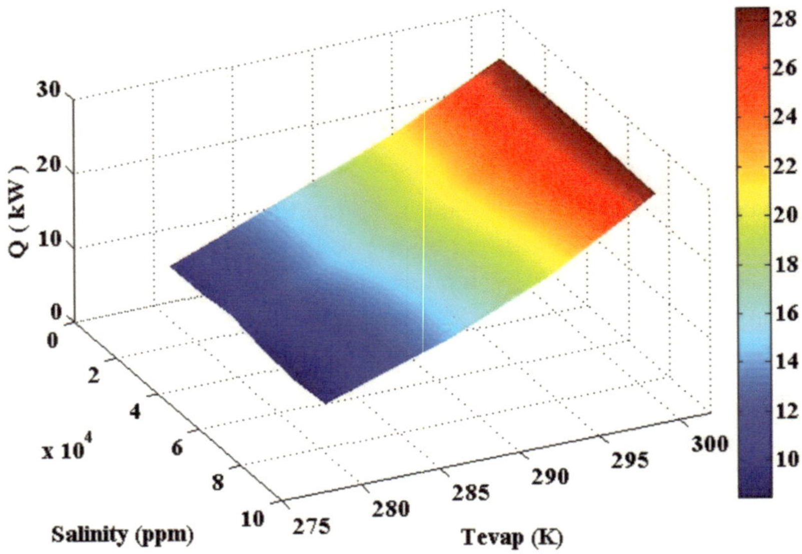

Figure 13. Effect of evaporator saturation temperature and feed salt concentration on heat input to evaporator.

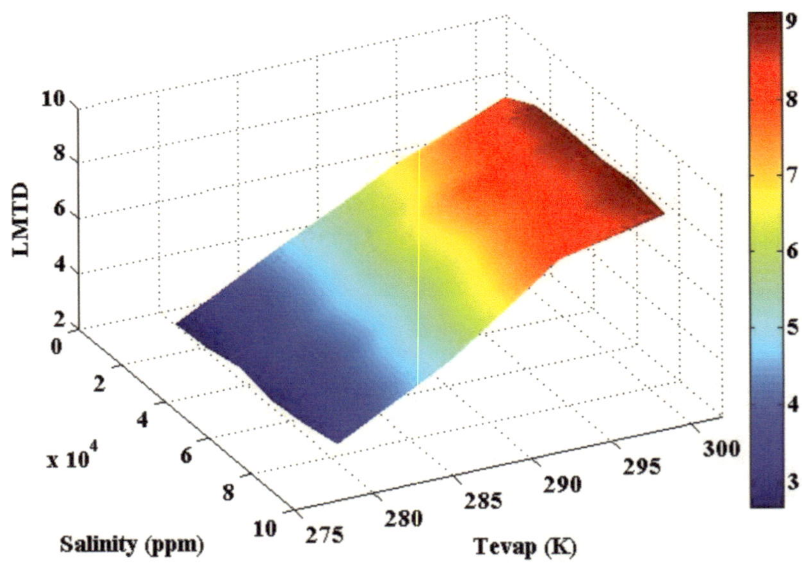

Figure 14. Effect of evaporator saturation temperature and feed salt concentration on LMTD.

6. Summary of chapter

Horizontal tube falling film evaporators can replace flooded and vertical tube evaporators because of their inherent advantages. Although horizontal falling film evaporators are advantageous, there is a lack of research data related to the heat transfer coefficient especially at low saturation temperatures less than 323 K. The heat transfer coefficient for low saturation temperature (typically in the zone of below ambient) and for a horizontal tube evaporator of special interest to desalination applications is essential.

Experiments are conducted to investigate the heat transfer coefficient for low saturation temperatures of 279–300 K corresponding to pressure ranges of 0.93–3.60 kPa. Salt concentration in the evaporator is investigated in the range of 15,000–90,000 ppm. The heat transfer coefficient calculated from experimental data is plotted for different salt concentrations.

At low saturation temperatures, below 298 K, the tendency for liquid film to flash into vapour is made easier by the rapid increase in the specific volume of vapour. For a given thermal gradient across the liquid film, the micro-bubble is readily generated at suitable nucleation sites, such as the grooved surfaces on the tubes. This conjecture of 'bubble-agitation boiling' is backed up by photographic evidence which indicates the presence of micro-bubble generation beneath the liquid layer. The effect of micro-bubble during film boiling reduces the thermal barrier within liquid film which is responsible for enhancement of heat transfer. At low saturation temperature, the evaporation is done by two mechanisms namely: thermally driven evaporation and bubble agitation-assisted evaporation. The basic domain of validation of traditional Han and Fletcher correlation is now extended through to capture the bubble-assisted evaporation. There is heat transfer enhancement due to bubble-assisted evaporation that increases the heat transfer coefficient value from two- to four-fold

A new falling film evaporation heat transfer coefficient is proposed with parameter regression including two basic mechanisms observed during experiments. The measured heat transfer coefficient from experimental data has uncertainty of less than 8%. The RMS error of regressed data is 3.5%. The effects of operational parameters namely salt concentration and saturation temperature on heat input and LMTD are also investigated. The proposed correlation can be used for the designing of low-pressure horizontal tubes falling film evaporators for process industry.

Nomenclature

μ_1 = Liquid viscosity (kg/m-sec)

k_1= Liquid conductivity (W/m K)

Pr = Prandtl number

S_o= Reference sea water salinity (30,000 ppm)

T_{evap} = Evaporator saturation temperature (K)

$T_{saturation}$ = Evaporator saturation temperature (K)

T_{ref}= Reference saturation temperature (K) (T_{ref} = 322.15 K)

$T_{ch,in}$ = Chilled water inlet temperature (K)

v_g = vapour specific volume (m^3/kg) (v_{ref} = 52. 65 m^3/kg at 295 K)

$\Delta T = T_{ch,in} - T_{evap}$

ρ_l = Liquid density (kg/m^3)

Re_Γ = Film Reynolds number

S = Feed water salinity (ppm)

q = input heat flux (W/m^2)

Abbreviations

EHTC	Evaporation heat transfer coefficient
FFEHTC	Falling film evaporation heat transfer coefficient
MED	Multi-effect desalination
MSF	Multi-stage flash evaporation
AD	Adsorption desalination
LMTD	Log mean temperature difference
Ppm	Part per million

Author details

Muhammad Wakil Shahzad*, Muhammad Burhan and Kim Choon Ng

*Address all correspondence to: muhammad.shahzad@kaust.edu.sa

Water Desalination and Reuse Center (WDRC), King Abdullah University of Science and Technology, Thuwal, Saudi Arabia

References

[1] Han JC, Fletcher LS. Falling film evaporation and boiling in circumferential and axial grooves on horizontal tubes. Industrial and Engineering Chemistry Process Design and Development. 1985;**24**(3):570-575

[2] Ribatski G, Jacobi AM. Falling film evaporation on horizontal tubes-a critical review. International Journal of Refrigeration. 2005;**28**(5):635-653

[3] Adib TA, Heyd B, Vasseur J. Experimental results and modeling of boiling heat transfer coefficients in falling film evaporator usable for evaporator design. Chemical Engineering and Processing. 2009;**48**(5):961-968

[4] Chun KR, Seban RA. Heat transfer to evaporating liquid films. Transactions of the ASME: Journal of Heat Transfer. 1971;**93C**(4):391-396

[5] Prost JS, Gonzaalez MT, Urbicain MJ. Determination and correlation of heat transfer coefficients in a falling film evaporator. Journal of Food Engineering. 2006;**73**(4):320-326

[6] Ahmed SY, Kaparthi R. Heat transfer studies of falling film heat exchangers. Indian Journal of Technology. 1963;**1**:377-381

[7] McAdams WH, Drew TB, Bays GS. Heat transfer to falling—water films. Transactions of ASME. 1940;**62**:627

[8] Herbert LS, Stern UJ. An experimental investigation of heat transfer to water in film flow. Canadian Journal of Chemical Engineering. 1968;**46**:401-407

[9] Uche J, Artal J, Serra L. Comparison of heat transfer coefficient correlations for thermal desalination units. Desalination. 2002;**152**(1-3):195-200

[10] Parken WH, Fletcher LS. An Experimental and Analytical Investigation of Heat Transfer to Thin Water Films on Horizontal Tubes. University of Virginia, Report UVA-526078-MAE, 1977, pp. 77-101

[11] Barba D, Felice RD. Heat transfer in turbulent flow on a horizontal tube falling film evaporator—a theoretical approach. Desalination. 1984;**51**:325-333

[12] Shah MM., A general correlation for heat transfer during film condensation inside pipes. International Journal of Heat and Mass Transfer. 1979;**22**:547-556

[13] Kutateladze SS. Fundamentals of Heat Transfer. New York: Academic Press; 1963

[14] Labuntsov DA. Heat transfer in film condensation of pure steam on vertical surfaces and horizontal tubes. Teploenergetika. 1957;**4**(7):72-79

[15] Fujita Y, Tsutsui M. Experimental investigation of falling film evaporation on horizontal tubes. Heat Transfer-Japanese Research. 1998;**27**:609-618

[16] Fujita Y, Tsutsui M, Zhou Z-Z. Evaporation heat transfer of falling films on horizontal tube - part 1, analytical study. Heat Transfer-Japanese Research. 1995;**24**:1-16

[17] Fujita Y, Tsutsui M, Zhou Z-Z. Evaporation heat transfer of falling films on horizontal tube - part 2, experimental stud. Heat Transfer-Japanese Research. 1995;**24**:17-31

[18] Liu ZH, Yi J. Falling film evaporation heat transfer of water/salt mixtures from roll-worked enhanced tubes and tube bundle. Applied Thermal Engineering. 2002;**22**(1):83-95

[19] Yang LP, Shen SQ. Experimental study of falling film evaporation heat transfer outside horizontal tubes. In: Conference on Desalination and the Environment Halkidiki, Greece; April 22-25, 2007. Desalination. 2008;**220**(1-3):654-660

[20] Parken WH, et al. Heat-transfer through falling film evaporation and boiling on horizontal tubes. Journal of Heat Transfer – Transactions of ASME. 1990;**112**(3)744-750

[21] Ribatski G, Thome JR. Experimental study on the onset of local dryout in an evaporating falling film on horizontal plain tubes. Experimental Thermal and Fluid Science. 2007;**31**(6):483-493

[22] Lorenz JJ, Yung D. Film breakdown and bundle-depth effects in horizontal tube, falling-film evaporators. Journal of Heat Transfer – Transactions of ASME. 1982;**104**(3):569-571

[23] Roques JF, Thome JR. Falling films on arrays of horizontal tubes with R-134a. Part II: Flow visualization, onset of dry out, and heat transfer predictions. Heat Transfer Engineering. 2007;**28**(5):415-434

[24] Aly G, Al-Hadda A, Abdel-Jawad M. Parametric study on falling film seawater desalination. Desalination. 1897;**65**:43-55

[25] Moeykens S, Pate MB. Spray evaporation heat transfer performance of R134a on plain tubes. Ashrae Transactions. 1994;**100**(2):173-184

[26] Moeykens S, Kelly JE, Pate MB. Spray evaporation heat transfer performance of R-123 in tube bundles. Ashrae Transactions. 1996;**102**(2):259-272

[27] Chang TB, Chiou JS. Spray evaporation heat transfer of R-141b on a horizontal tube bundle. International Journal of Heat and Mass Transfer. 1998;**42**:1467-1478

[28] Moeykens S, Pate MB. The effects of nozzle height and orifice size on spray evaporation heat transfer performance for a low-finned, triangular-pitch tube bundles with R-134a. Ashrae Transactions 1995;**101**(2):420-433

[29] Moeykens S, Pate MB. Effect of lubricant on spray evaporation heat transfer performance of R-134a and R-22 in tube bundles. Ashrae Transactions. 1996;**102**(1):410-426

[30] Moeykens S, Newton BJ, Pate MB. Effects of surface enhancement, film-feed supply rate, and bundle geometry on spray evaporation heat transfer performance. Ashrae Transactions. 1995;**101**(2):408-419

[31] Bourouni K, Martin R, Tadrist L, Tadrist H. Modelling of heat and mass Transfer in a horizontal tube falling film evaporators for water desalination. Desalination. 1998;**116**: 165-184

[32] Yang L, Shen S. Experimental Study of Falling Film Evaporation Heat Transfer outside Horizontal Tube. Desalination. 2008;**220**:654-660

[33] Xu L, Ge M, Wang S,Wang Y. Heat Transfer Film Coefficients of Falling Film Horizontal Tube Evaporators. Desalination. 2004;**166**:223-230

[34] Yang L, Shen S. Experimental study of falling film evaporation heat transfer outside horizontal tubes. Desalination. 2008;**220**:654-660

[35] Chun KR, Seban RA. Heat Transfer to evaporating liquid films. ASME Journal of Heat Transfer. 1971;**11**:391-396

[36] Alhusseini AA, Tuzla K, Chen JC. Falling film evaporation of single component liquids. International Journal of Heat and Mass Transfer. 1998;**41**(12):1623-1632

[37] Shmerler JA, Mudawwar I. Local evaporative heat transfer coefficient in turbulent free-falling liquid films. International Journal of Heat and Mass Transfer. 1988;**31**(4):731-742

[38] Chien LH, Tsai YL. An experimental study of pool boiling and falling film evaporation on horizontal tubes in R-245fa. Applied Thermal Engineering. 2011;**31**:4044-4054

Chapter 9

Application of Taguchi-Based Design of Experiments for Industrial Chemical Processes

Rahul Davis and Pretesh John

Additional information is available at the end of the chapter

http://dx.doi.org/10.5772/intechopen.69501

Abstract

Design of experiment is the method, which is used at a very large scale to study the experimentations of industrial processes. It is a statically approach where we develop the mathematical models through experimental trial runs to predict the possible output on the basis of the given input data or parameters. The aim of this chapter is to stimulate the engineering community to apply Taguchi technique to experimentation, the design of experiments, and to tackle quality problems in industrial chemical processes that they deal with. Based on years of research and applications, Dr. G. Taguchi has standardized the methods for each of these DOE application steps. Thus, DOE using Taguchi approach has become a much more attractive tool to practicing engineers and scientists. And since the last four decades, there were limitations when conventional experimental design techniques were applied to industrial experimentation. And Taguchi, also known as orthogonal array design, adds a new dimension to conventional experimental design. Taguchi method is a broadly accepted method of DOE, which has proven in producing high-quality products at subsequently low cost.

Keywords: DOE, Taguchi method, industrial chemical processes, parameter optimization, ANOVA

1. Introduction

Industries are engaged in a variety of activities such as developing new products, improving previous designs, maintenance, controlling and improving ongoing processes and some more. Experimentation is a frequent task in these activities to measure and analyse the output, and for this purpose engineers/researchers use many tools like statistics, analytical models, etc., regardless of their background in it [1].

Montgomery [2] writes,

> *"Experiments are performed in almost any field of enquiry and are used to study the performance of processes and systems. The process is a combination of machines, methods, people and other resources that transforms some input into an output that has one or more observable responses. Some of the process variables are controllable, whereas other variables are uncontrollable, although they may be controllable for the purpose of a test. The objectives of the experiment include: determining which variables are most influential on the response, determining where to set the influential controllable variables so that the response is almost always near the desired optimal value, so that the variability in the response is small, so that the effect of uncontrollable variables are minimized."*

In today's era, the purpose of experiments in industries is essentially optimization and robust design analysis (RDA, which is used to make the system less sensitive to variations in uncontrollable noise factors or in other words to make the system robust). DOE, or experimental design, is the name given to the techniques used for guiding the choice of the experiments to be performed in an efficient way. In a general way, the process analysis can be expressed as the study of the cause-effect relationships which may be carried out by drawing inferences from a finite number of samples. And one of the most important purposes of it is to design sampling experiments that are productive and cost-effective and provide a sufficient data base in a qualitative sense [3]. Design of experiments has been applied successfully in diverse fields such as agriculture (improved crop yields have created grain surpluses), the petrochemical industry (for highly efficient oil refineries) and Japanese automobile manufacturing (giving them a large market share for their vehicles), and still its implementation area is spreading and providing the optimized results. These developments are due in part to the successful implementation of design of experiments. The reason to use design of experiments is to implement valid and efficient experiments that will produce quantitative results and support sound decision-making [4].

1.1. Brief history

Statistical experimental design, together with the basic ideas underlying DOE, was born in the 1920s from the work of Sir Ronald Aylmer Fisher [5]. Fisher was the statistician who created the foundations for modern statistical science. The second era for statistical experimental design began in 1951 with the work of Box and Wilson [6], who applied the idea to industrial experiments and developed the response surface methodology (RSM), which is used to find out the relationships between various process parameters and one or more responses. The work of Dr. Genichi Taguchi in the 1980s [7], despite having been very controversial (described briefly in heading 2.4), had a significant impact in making statistical experimental design popular and stressed the importance it can have in terms of quality improvement.

Usually, data subject to experimental error (noise) are involved, and the results can be significantly affected by noise. Thus, it is better to analyse the data with appropriate statistical methods. The basic principles of statistical methods in experimental design are replication, randomization and blocking. Replication is the repetition of the experiment to obtain a more precise result (sample mean value) and to estimate the experimental error (sample standard deviation). Randomization refers to the random order in which the runs of the experiment are to be performed. In this way, the conditions in one run neither depend on the conditions

http://dx.doi.org/10.5772/intechopen.69501

of the previous run nor predict the conditions in the subsequent runs. Blocking aims at isolating a known systematic bias effect and prevents it from obscuring the main effects [8]. This is achieved by arranging the experiments in groups that are like one another. In this way, the sources of variability are reduced, and the precision is improved.

The design of experiments (DOE) is explained by Lye [9], as a methodology for systematically applying statistics to experimentation. In DOE, a sequence of tests is designed in which purposeful vary the input parameters (factors) of a product or process to examine the reasons of the variation in the output response [10]. By the end of the twentieth century, DOE was no longer viewed as merely a stand-alone tool, because it was packaged together with a structured initiative for business improvement known as Six Sigma. Moreover, an increased emphasis on DOE took place during this period in Six Sigma literature [11]. DOE is a good tool to understand and optimize products or process parameters. It is quick as well as cost-effective.

1.2. Advantages of DOE

With real engineering examples, Czitrom [12] listed the following advantages of DOE:

- A good amount of data can be obtained with lesser resources (experiments, time, material, etc.).
- The estimates of the effect of each factor (variable) on the response are more precise.
- It is a systematic way to estimate the interactions between the process factors.
- There is experimental information in a larger region of the factor space.

1.3. DOE techniques

A survey was carried out within the industry which identifies the needs of using an efficient and practical technique for the experimentation. It was surveyed that 76% of industries consider themselves in need of a methodology [13]. So here are listing some of the techniques that are in use in Industries. The list of the techniques considered is far from being complete since the aim of the section is just to introduce the reader into the topic showing the main techniques which are used in practice [14].

- Randomized complete block design
- Latin square
- Full factorial
- Fractional factorial
- Central composite
- Box-Behnken [15]
- Plackett-Burman [16]
- Taguchi [7]

- Random
- Halton, Faure and Sobol sequences
- Latin hypercube
- Optimal design
- Response surface design

Several DOE techniques are available to the experimental designer. However, as it always happens in optimization, there is no best choice. The correct DOE technique selection depends on the problem to be investigated and on the aim of the experimentation.

M. Cavzzuti [14] concluded that items to be considered are:

a. The number of experiments N which can be afforded. In determining the number of experiments, an important issue is the time required for a single experiment. There is a lot of difference between whether the response variable is extracted from a quick simulation in which a number is computed or taken from a spreadsheet or it involves the setting up of a complex laboratory experiment. In the former case, it could take a fraction of a second to obtain a response, and in the latter one, each experiment could take days.

b. The number of parameters k of the experiment. For many DOE techniques, the number of experiments required grows exponentially with the number of parameters (**Figure 1**). Not

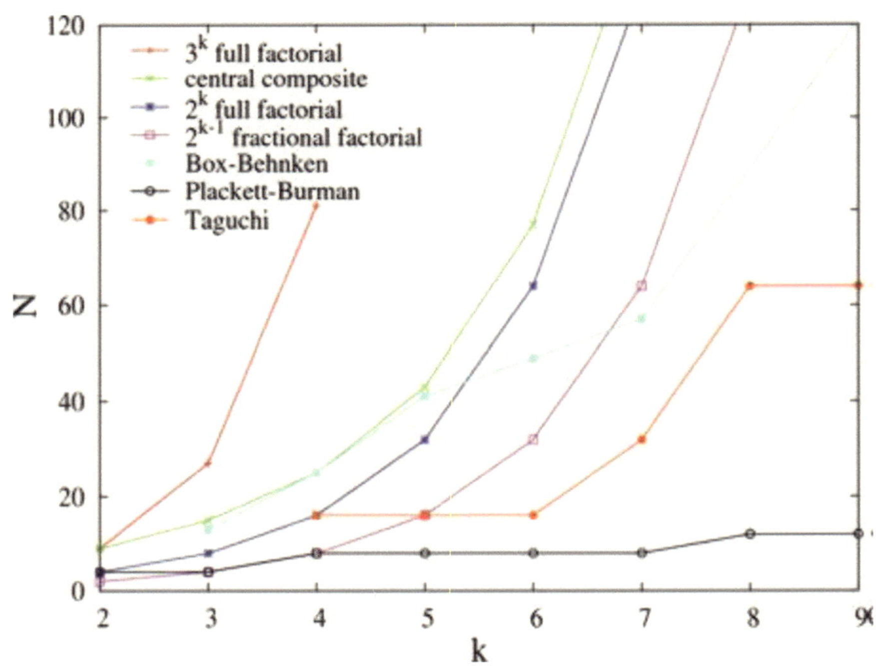

Figure 1. Number of experiments required by the DOE techniques [14].

necessarily to use a cheap technique is the best choice, because a cheap technique means imprecise results and insufficient design space exploration. Unless the number of experiments which can be afforded is high, it is important to limit the number of parameters as much as possible to reduce the size of the problem and the effort required to solve it. Of course, the choice of the parameters to be discarded can be a particularly delicate issue. This could have done by applying a cheap technique (like Plackett-Burman etc.) as a preliminary study for estimating the main effects.

2. Introduction to Taguchi method

2.1. Brief history

After the Second World War, allied forces observed some of the major drawbacks of the Japanese telephone system, that is, extremely poor quality and unsuitability for long-term communication purposes. To overcome these drawbacks, an improved system was required, for this the allied command recommended establishing research facilities to develop a state-of-the-art communication system. At that time, the electrical communication laboratories (ECL) were came on the stage with Dr. Genichi Taguchi (**Figure 2**) in charge of improving the R&D productivity and enhancing product quality. It was observed that the ratio of the time and money expended on engineering experimentation and testing is very high than the efforts given to the process of creative brainstorming to minimize the expenditure of resources. He noticed that the process of inspection, screening and salvaging cannot improve poor quality. The inspection process is done to check the quality but it can't increase the quality by itself. Therefore, he believed that quality concepts should be based upon, and developed around, the philosophy of prevention.

This moved Taguchi to develop new optimizing methods of the processes of engineering experimentation. He believed that the best way to improve quality was to design and build it

Figure 2. Dr. G. Taguchi [17].

into the product. He quoted that [18], *"Cost is more important than quality but quality is the best way to reduce cost."*

He developed the techniques which are now known as Taguchi methods (TM). His main contribution lies not in the mathematical formulation of the design of experiments, but rather in the accompanying philosophy. Taguchi method is different from the traditional techniques because of Taguchi's concepts of design. By his methods, he developed Robust Manufacturing systems that are insensitive to daily or seasonal variations of environment, wears or other noise factors. His philosophy had far reaching consequences, yet it is founded on three very simple concepts [19].

Taguchi's new technique consist of three concepts about quality, these are:

1. Quality should be designed into the product and not inspected into it.
2. Quality is better achieved by minimizing the deviation from a target. The product should be so designed that it is immune to uncontrollable environmental factors.
3. The cost quality should be measured as a function of deviation from the standard and the losses should be measured system-wide.

In Taguchi's thinking, the quality improvement is an ongoing effort. He endeavoured continually to reduce the variation around the target value. Right population selection as near to the target value or desired value is the first stair step of the quality improvement. And to accomplish this, Taguchi designed experiments using especially constructed tables known as "orthogonal arrays" (OA). It makes the design of experiments very easy and consistent.

Taguchi's two most important contributions to quality engineering are as follows:

i) The use of Gauss's quadratic loss function to quantify quality.

ii) The development of robust designs (parameter and tolerance design).

Since the early 1980s when applications to different industries began in western hemisphere, the Taguchi method is evaluated in different platforms like books, articles, panel discussions, etc. Taguchi methods have touched most of the manufacturing processes to optimize the process in such a way so that noise factors do not affect the output. Several reports [20–29] evaluated Taguchi methods from a statistical standpoint. In these reports, the parameter design received the most attention. These reports confirm that Dr. Taguchi made important contributions to quality engineering; however, without some basic statistical knowledge, it is hard to apply his technique. Specifically, the use of signal-to-noise ratios in identifying the nearly best factor levels to minimize quality losses may not be efficient [30].

2.2. Taguchi method

In Taguchi method, we assume that we are designing an engineering system—it might be a machine that performs some intended function or it might be a production process. We use the knowledge of fundamental about the system and process parameters for efficient experimentation.

We can skip all the extra effort that might have gone in to investigating interactions. By this, we can decrease the number of factors. Taguchi categorize the factors in two sets:

1. Control factors, which are under our control.
2. Noise factors, which are not under our control, except during experiments in the laboratories.

In the 1920, Sir R. A. Fisher first proposed the DOE with multiple factors known as Factorial Design of Experiments. In the full factorial design, we work on the all possible combinations extruded from the preselected set of factors. Mostly all industrial experiments depends upon on the number of factors and if we consider all possible combination then it becomes harder and also time consuming to execute a large sequence of experiments. The full fractional design consists of k^n experiments, where "k" is the number of levels of factors and "n" is the number of factors (factors are the different variables, which determine the functionality or performance of a product or system). After that, partial factorial method came into the existence. In this, the number of experiments is reduced, and only a small set from all possibilities is selected, which produces the most information.

Taguchi's approach complements the following two important areas:

- Taguchi constructed a special set of orthogonal arrays (OA) to lay out his experiments. Taguchi prepared a new set of standard OAs which could be used for a number of experimental situations.
- He also devised a standard method for analysis of the results.

As mentioned, the full factorial design requires many experiments to be carried out. It becomes laborious and complex, if the number of factors increases. To overcome this problem, Taguchi suggested a specially designed method called the use of orthogonal array. In this, we can study the more factors or parameter space with lesser number of experiments. Opposite to full factorial analysis, the Taguchi method reduces the number of experimental runs to a reasonable one, in terms of cost and time, using orthogonal arrays [31]. For example, if there are three factors called A, B and C, all are examined with two levels called "1" and "2" (in general, they are referred as "1" and "−1"). Then, according to the full fractional, the number of experiments should be $2^3 = 8$. In **Table 1**, the full fractional array is shown; such experiments can find all main and two and three factor interactions.

But at the same time, Taguchi's L_8 array can deal with seven factors and their two levels as shown in **Table 2**, such experiments can find all seven main factors effects.

So, it can be seen that the full fractional method is dealing with three factors, and orthogonal array is dealing with seven factors in same eight experimental runs.

Taguchi method is based on mixed levels, highly fractional factorial designs and other orthogonal designs. It distinguishes between control variables and noise variables. In this, we choose two sets of parameters, that is, controlled and noise parameters or variables. Respectively, we choose orthogonal designs also. The design chosen for the controlled variable is known as inner array and for the noise variables is known as outer array. The combination of the inner

A	B	C	AB	BC	AC	ABC
1	1	1	1	1	1	1
2	1	1	2	1	2	2
1	2	1	2	2	1	2
2	2	1	1	2	2	1
1	1	2	1	2	2	2
2	1	2	2	2	1	1
1	2	2	2	1	2	1
2	2	2	1	1	1	2

Table 1. Full fractional factor assignments to experimental array columns.

A	B	C	D	E	F	G
1	1	1	1	1	1	1
2	1	1	2	1	2	2
1	2	1	2	2	1	2
2	2	1	1	2	2	1
1	1	2	1	2	2	2
2	1	2	2	2	1	1
1	2	2	2	1	2	1
2	2	2	1	1	1	2

Table 2. Orthogonal array factor assignments to experimental array columns.

and the outer arrays gives the crossed array, which is the list of all the samples scheduled by the Taguchi method. It means that for each sample in the inner array, the full set of experiments of the outer array is performed. The advantage of this cross combination is that it provides the information about the relation between the parameters which plays very important for the robust system designing.

Then, it is recommended by Dr. Taguchi to use the quality loss function to measure the performance characteristics. The quality loss function is a continuous function that is defined in terms of the deviation of a design parameter from an ideal or target value. The value of this loss function is further transformed into signal-to-noise (S/N) ratio. Performance characteristics are available in three categories to determine the S/N ratio:

- Nominal-the-Best
- Larger-the-Best
- Smaller-the-Best

The Taguchi method has four basic phases in the optimization process, these are as follows:

1. First phase is to timely thinking about the quality characteristics and determining the parameters which important to the product or process.
2. In second phase the experiments sequence is designed and executed accordingly.
3. In third phase of the optimization process the statistical analysis is done to determine the optimum conditions.
4. Finally in the fourth phase the confirmation test is run with the optimum conditions.

Barrado et al. [32] expanded above-mentioned four phases into the following steps for implementing Taguchi experimental design:

Step 1. Selection of the output or target parameters.

Step 2. Identification of the input parameters and their levels.

Step 3. Determining the suitable orthogonal array (OA).

Step 4. Assign factors and interactions to the columns of the array.

Step 5. Conduct the experiments.

Step 6. Statistical analysis and the signal-to-noise ratio and determining the optimum setting of the factor levels.

Step 7. Perform confirmatory experiment (if necessary).

First of all, the input parameters and the respective operation levels and output or responses are selected. After this choice, the best fitted or economical matrix experiment or orthogonal array is selected [33, 34]. Here, in **Table 3**, the array selector is shown.

After performing the experiments as per the chosen array, we choose the desired signal-to-noise ratio function (smaller-the-better, larger-the-better and nominal-the-better). The S/N ratio is a logarithmic function which can also be defined as an inverse of variance. Generally in the optimization of the process or product design and in minimizing the variability we use it. If we maximize the S/N ratio, we reduce the variability of the process against undesirable changes in noise factors. Because S/N ration and variance are inversely proportional, so the chosen factors should produce maximum value of S/N so that we get the minimum variability.

Three types of common problems and respective signal-to-ratio function are presented in **Table 4**.

In **Table 4**, y_i denotes the nth observations of response variable and μ^2 denotes the square of mean and σ^2 the variance of the observations of the replicated response values.

For analysing the data, one of the most common methods used is ANOVA. ANOVA is a statistical technique that assesses potential differences in a scale-level dependent variable by a nominal-level variable having two or more categories [35]. The ANOVA is developed by Sir

Numbers of Parameters (P)	Number of levels			
	2	3	4	5
2	L4	L9	L16	L25
3	L4	L9	L16	L25
4	L8	L9	L16	L25
5	L8	L18	L16	L25
6	L8	L18	L32	L25
7	L8	L18	L32	L50
8	L12	L18	L32	L50
9	L12	L18	L32	L50
10	L12	L27	L32	L50
11	L12	L27		L50
12	L16	L27		L50
13	L16	L27		
14	L16	L36		
15	L16	L36		
16	L32	L36		
17	L32	L36		
18	L32	L36		
19	L32	L36		
20	L32	L36		
21	L32	L36		
22	L32	L36		
23	L32	L36		
24	L32			
25	L32			
26	L32			
27	L32			
28	L32			
29	L32			
30	L32			
31	L32			

Table 3. Taguchi orthogonal array selector.

Choose...	S/N ratio formulas	Use when the goal is to...
Smaller the better	$S/N = -10\log\left[\frac{1}{n}\sum_{i=1}^{n} y_i^{\ 2}\right]$	Minimize the response
Nominal the better	$S/N = 10\log\frac{\mu^2}{\sigma^2}$	Target the response and you want to base the S/N ratio on means and standard deviations
Larger the better	$S/N = -10\log\left[\frac{1}{n}\sum_{i=1}^{n}\frac{1}{y_i^{\ 2}}\right]$	Maximize the response

Table 4. Types of problems and respective signal-to-ratio function.

Ronald Fisher in 1918. It is the extended version of T test and Z test. The reason of its origin was the limitations of the T & Z tests which have the problem of only allowing the nominal level variable to have just two categories. This analysis method is also famous with the title "The Fisher Analysis of Variance."

The use of ANOVA depends on the research design. Commonly, ANOVAs are used in three ways:

a. One-way ANOVA

b. Two-way ANOVA

c. N-way multivariate ANOVA.

2.3. Applications of Taguchi method in industrial chemical processes

The Taguchi method is used whenever the settings of interest parameters are necessary, not only for manufacturing processes. Therefore, the Taguchi approach is used in many domains such as environmental sciences [36, 37], agricultural sciences [38], physics [39], statistics [40], management and business [41], medicine [42] and in chemical processes as well [43]. Here, some of the literature reviews are given, which clearly show the application of Taguchi method in the chemical processes of various industries.

The identification and incorporation of quality costs and robustness criteria are becoming a critical issue while addressing chemical process design problems under uncertainty. Fernando P. Bernardo *et al.* [44] used Taguchi loss functions along with other robustness criteria and presented a systematic design framework. They conducted their study within a single-level stochastic optimization formulation, and an efficient cubature technique is used for the estimation of the expected values. An optimal design was discovered, together with a robust operating policy. It was observed that these parameters maximizes average process performance.

Kundu et al. [45] investigated the optimal operation conditions to prepare activated carbon (AC) using palm kernel shell (PKS). They choose four control factors for their research which were irradiation time, microwave power, concentration of impregnation substance which was phosphoric acid and impregnation ratio between acid and PKS aided by Taguchi optimization method. After successful implementation of Taguchi method, the optimal settings were found and the optimal levels were microwave power of 800 W, irradiation time of 17 min, impregnation ratio of 2 and concentration of acid 85% (undiluted). After the confirmation test with optimal settings, activated carbon with high BET surface area of 1473.55 $m^2\ g^{-1}$ and high porosity was obtained.

Zolfaghari et al. [46] presented a systematic optimization approach by using the Taguchi method for removal of lead (Pb) and mercury (Hg) by a nanostructure, zinc oxide-modified mesoporous carbon CMK-3 denoted as Zn-OCMK-3. CMK-3 was synthesized by using SBA-15 and then oxidized by nitric acid. Their investigation using Zn-OCMK-3 in the frame of the Taguchi method brought forth the optimum conditions for Pb and Hg adsorption from water. The determined optimum conditions for removal of Pb (II) and Hg (II) were the agitation time of 120 min, the initial concentration of 10 mg/l, the temperature of 35 ◦C, the dose of 0.7 g/l, and the pH of 6. Based on the confirmation test under optimum conditions for Zn-OCMK-3, it was observed that this nanoporous carbon is very effective in removing the lead and mercury with a high pollutant removal efficiency (PRE) i.e. 97.25% for Pb (II) and 99% for Hg (II). Removal of lead and mercury were highly concentration dependent. Number of Pb and Hg ions highly increase from initial concentration of 10–400 mg/l. It is also observed that a lot of ions cannot be adsorbed at high concentration, which reduce the removal efficiency.

Venkata Mohan et al. [47] applied design of experimental methodology (DOE) by Taguchi approach to find out the effects of selected factors on the H_2 production with the final aim of optimizing the process. They selected four factors for their research study, that is, inlet pH, inoculum type, feed composition and inoculum pre-treatment methods. Here, also Taguchi method enhanced the process. Results showed significant variation and process efficiency. Among the factors studied with respect to H_2 production, feed composition showed stronger influence followed by inlet pH, pre-treatment method and origin of the inoculum. Taguchi approach also permits process optimization of system by a set of independent factors (over a specific region of interest (levels) by identifying the influence of individual factors, establishing the relationship between variables and operational conditions and finally estimate the performance at optimum levels obtained. By using optimized conditions of the factors, the rate of H_2 production can be enhanced by almost threefold (0.376–1.166 mmol/day), same positive enhanced results were recorded for substrate degradation also.

Taguchi-based DOE methodology provides a systematic and efficient mathematical approach to understand complex process of fermentative H_2 production and substrate degradation for the optimization of the near optimum design parameters, only with a few well-defined experimental sets [48].

Messias Borges Silva *et al.* [49] applied Taguchi method in optimization of the treatment conditions of polyester-resin effluent by means of Advanced Oxidative Processes (AOPs). The output parameter were Chemical oxygen demand (COD). Their study consist of two type of evaluation. First one is, the influence individual independent parameter and the second one is the respective interactions effect. This allowed determining the most statistically significant variables and the optimal conditions. The treatment by means of advanced oxidative process provided an approximate 35% reduction in chemical oxygen demand of the polyester-industry wastewater. However, when compared to studies describing the treatment of this effluent by Advanced Oxidative Processes, it was seen that the results were relevant. This investigation confirms the effectiveness of Taguchi Method in the industrial chemical processes.

Liao et al. [50] applied the Taguchi method and designs of experiments (DOE) approach to optimize parameters for chemical mechanical polishing (CMP) processes in wafer manufacturing. Planning of experiments was based on a Taguchi orthogonal array table to determine an optimal setting, and significant results were found.

S. V. Mohan *et al.* [51] applied the Taguchi robust experimental design (DOE) methodology on a dynamic anaerobic process treating complex wastewater by an anaerobic sequencing batch biofilm reactor (AnSBBR). Their work was to optimize the process and also at the other hand to evaluate the influence of distinct factors on the process. They considered the uncontrollable (noise) factors. This is the first kind of study of anaerobic process evaluation and process optimization by using the Taguchi methodology adopting dynamic approach.

The biological oxidation of ferrous ion by iron-oxidizing bacteria is potentially a useful industrial process for removal of H2S from industrial gases, desulphurization of coal, removal of sulfur dioxide from flue gas, treatment of acid mine drainage and regeneration of an oxidant agent in hydrometallurgical leaching operations that pH of feed solution has the most contribution in the biooxidation rate of ferrous ion. When the parameters were set according to Taguchi approach, the biological reaction rate was obtained. Mousavi et al. [52] studied and find out the optimum values of the process parameters on the ferrous biooxidation rate by immobilization of a native Sulfobacillus species on the surface of low-density polyethylene (LDPE) particles in a packed-bed bioreactor using Taguchi method. L_{16} array was used with five factors and their four levels. Temperature, initial pH of feed solution, dilution rate and initial concentration of Fe^{3+} and aeration rate are considered as input parameters in Taguchi technique. Analysis of variance (ANOVA) was used to determine the optimum conditions and most significant process parameters affecting the reaction rate. Results indicated that pH of feed solution has the most contribution in the biooxidation rate of ferrous ion. When the parameters were set according to Taguchi approach, the biological reaction rate was obtained 8.4 g L^{-1} h^{-1}.

Taguchi robust design method with L_9 orthogonal array was implemented to optimize experimental conditions for the preparation of nano-sized silver particles using chemical reduction method [53]. The parameters for chemical-mechanical polishing (CMP) in an ultra-large-scale integrated (ULSI) planarization process are explored using the Taguchi method [54]. So, it can be seen easily the popularity of the Taguchi method in industrial chemical processes.

2.4. Advantages and disadvantages of Taguchi method

The foundation of the DOE in Taguchi method (TM) is orthogonal array design that is very simple method for analysing the outputs. His work was controversial and criticized by some researchers for being inefficient and ineffective in some cases [55]. But still the simplicity of Taguchi method increased its use in manufacturing industries. Here, some of the advantages and disadvantages are discussed briefly which clears its popularity even in the clouds of controversies. A survey shows that 51% of respondents are familiar with TM, although only 14% of them apply it [13].

The key step of the TM is to increasing the quality level with less affecting the cost factor. TM provides the optimal settings for the processes which can improve quality, and these settings attained from TM are also insensitive to the variation of noise factors. Basically, classical process parameter design is complex and not easy-to-use, and at the other hand, Taguchi technique is user friendly [56].

Another advantage of the Taguchi method is that it emphasizes a mean performance characteristic value close to the target value rather than a value within certain specification limits, thus improving the product quality. Also, it is straightforward and easy to apply to many engineering situations, this property makes it a powerful yet simple tool for industries. It can be used to quickly narrow the scope of a research project or to identify problems in a manufacturing process from data already in existence [57].

It is probably unfortunate that the important concepts advocated by Taguchi have been overshadowed by controversy associated with his approach to modelling and data analysis. There have been several research papers and books which explain, review, or raise questions on the Taguchi's ideas [55, 58]. One of its demerit is that the results obtained from it are only relative. It is unable to indicate exactly which parameters have the highest effect on the performance or response.

Another debating topic is that when orthogonal array do not tests all the possible combinations of the factors, then this method should not be adopted with all relationships between all factors. The Taguchi method has been criticized for its difficulty in accounting for interactions between parameters.

Another limitation is that the Taguchi methods are offline. So in the case of dynamically changing processes for example a simulation study, the TM is not appropriate. And also, the Taguchi method is most effectively applied at the designing stage of a product development, it cannot help after the process is started. Because TM deals with designing quality rather than correcting for poor quality [59].

3. Conclusion

Industries are in the need of the method of conducting experiments which optimize processes and increase the quality of products. Same is desired in chemical industrial processes also. For this, design of experiments is the basic step of quality improvement via optimized processes. And this requires proper planning and layout of the experiments and accurate analysis of the results. And Dr. G. Taguchi studied these issues very well and developed his method. Thus, DOE using Taguchi approach has become a much more attractive tool to practicing engineers and scientists. Since the conventional experimental design techniques were applied to industrial applications or processes, these conventional methods were always covered with drawbacks and limitations. And Taguchi array, also known as orthogonal array design, adds a new dimension to conventional experimental design. Taguchi method is a broadly accepted method of DOE which has proven in producing high-quality products at subsequently low cost. In most of the industrial applications or processes, the researchers and scientists used Taguchi method with other analytical tools in their research works, and in industrial chemical processes, it is also showing great results in optimization of the processes. A very important and fundamental part of Taguchi's method is to make sure that the product performs well even in noise; it helps in making the product long lasting. Taguchi's method can be applied in a very short amount of time; it does not take a lot of effort and improves the processes dramatically. The manufacturing industries can improve their processes very quickly and efficiently by applying the Taguchi's method. Around Industrial chemical processes, Taguchi method is showing the outstanding performance by optimizing the process parameters and reducing the number of experiments via orthogonal arrays. Taguchi method gives a new height to the processes by making them cost-effective and quick with improved quality.

Author details

Rahul Davis* and Pretesh John

*Address all correspondence to: rahuldavis2012@gmail.com

Department of Mechanical Engineering, Shepherd Institution of Engineering and Technology, SHUATS, Allahabad, India

References

[1] Bisgaard S. Teaching statistics to engineers. The American Statistician. 1991;**45**(4):274-283. DOI: 10.2307/2684452

[2] Montgomery DC. Design and Analysis of Experiments. 5th ed. New York: Wiley; 2000

[3] Toutenburg H, Shalabh. Statistical Analysis of Designed Experiments. 3rd ed. Heidelberg: Springer Texts in Statistics; 2009. DOI 10.1007/978-1-4419-1148-3_1

[4] Telford JK. A brief introduction to design of experiments. Johns Hopkins APL Technical Digest. 2009;**27**(3):224-232

[5] Fisher RA. Statistical Methods for Research Workers. Edinburgh: Oliver and Boyd; 1925

[6] Box GEP, Wilson KB. Experimental attainment of optimum conditions. Journal of the Royal Statistical Society. 1951;**13**:1-45

[7] Taguchi G, Wu Y. Introduction to Off-line Quality Control. Nagoya: Central Japan Quality Control Association; 1980

[8] NIST/SEMATECH. NIST/SEMATECH e-handbook of statistical methods. 2006. Retrieved from http://www.itl.nist.gov/div898/handbook/

[9] Lye LM. Tools and toys for teaching design of experiments methodology. In: 33rd Annual General Conference of the Canadian Society for Civil Engineering. Toronto, Ontario, Canada. 2005

[10] Montgomery DC. Design and Analysis of Experiments. 6th ed. John Wiley & Sons, Inc; Hoboken, NJ. 2005

[11] Brady JE, Allen TT. Six Sigma literature: A review and agenda for future research. Quality and Reliability Engineering International. 2006;**22**:335-367

[12] Czitrom V. One factor at a time versus designed experiments. The American Statistician. 1991;**53**(2):126-131

[13] Tanco M, Viles E, Ilzarbe L, Álvarez MJ. Manufacturing Industries Need Design of Experiments (DOE). In Proceedings of the World Congress on Engineering. 2007;**2**:1108-1112. London: Newswood Limited. ISBN:978-988-98671-2-6

[14] Cavazzuti M. Design of Experiments. In Optimization Methods: From Theory to Design. Springer-Verlag Berlin Heidelberg. 2013, pp. 13-42. DOI:10.1007/978-3-642-31187-1_2

[15] Box GEP, Behnken D. Some new three level designs for the study of quantitative variables. Technometrics. 1960;**2**:455-475

[16] Plackett RL, Burman JP. The design of optimum multifactorial experiments. Biometrika. 1946;**33**(4):305-325

[17] http://wiki.mbalib.com/w/images/5/5a/%E7%94%B0%E5%8F%A3%E7%8E%84%E4%B8%80.jpg

[18] AZquotes. Taguchi's quote. Retrieved from http://www.azquotes.com/quote/641638. Accessed on January 2017

[19] Ranjit R. A Primer on the Taguchi Method (Competitive Manufacturing Series), Van Nostrand Reinhold. ISBN-10: 0442237294, ISBN-13: 978-0442237295

[20] Box GEP, Fung CA. Minimizing Transmitted Variation by Parameter Design. Center for Quality and Productivity Improvement, University of Wisconsin-Madison, (Report no. 8). 1986

[21] Box GEP, Bisgaard S, Fung C. An explanation and critique of Taguchi's contributions to quality engineering. Quality and Reliability Engineering International. 1988;**4**(2):123-131

[22] Box GEP, Jones S. Designing products that are robust to the environment. Total Quality Management. March 1992;**3**(3):265-282

[23] Bisgaard S. Quality Engineering and Taguchi Methods: A Perspective. The Center for Quality and Productivity Improvement of University of Wisconsin, Madison (CQPI Report no. 40). 1990

[24] Czitrom V. An Application of Taguchi's Methods Reconsidered. ASA Meeting, Washington, DC. 1946

[25] Bisgaard S, Diamond NT. An Analysis of Taguchi's Method of Confirmatory Trials. CQPI Reports (Report no. 60). 1990

[26] Bisgaard S. Process Optimization- going Beyond Taguchi Methods. CQPI Reports (Report no. 70). 1991

[27] Bisgaard S. A comparative analyses of the performance of Taguchi's linear graphs. CQPI Reports (Report no. 82). 1992

[28] Bisgaard S, Ankenman B. Analytic Parameter Design. CQPI Reports (Report no. 103). 1993

[29] Steinberg D, Burnsztyn D. Noise Factors, Dispersion Effects and Robust Design. CQPI Reports (Report no. 107). 1946

[30] Maghsoodloo S, Ozdemir G, Jordan V, Huang C H. Strengths and limitations of Taguchi's contributions to quality, manufacturing, and process engineering. Journal of Manufacturing Systems. 2004;**23**(2):73-126. DOI: 10.1016/S0278-6125(05)00004-X

[31] Taguchi G, Jugulum R, Taguchi S. Computer-based Robust Engineering: Essentials for DFSS. Milwoukee, WI: ASQ Quality Press; 2004

[32] Barrado E, Vega M, Grande P, Del Valle JL. Optimization of a purification method for metal-containing wastewater by use of a Taguchi experimental design. Water Research. 1996;**30**(10):2309-2314. DOI: 10.1016/0043-1354(96)00119-4

[33] Padke MS. Quality Engineering Using Robust Design. Englewood Cliffs: Prentice Hall; 1989

[34] Taguchi G, Konishi S. Taguchi Methods Orthogonal Arrays and Linear Graphs: Tools for Quality Engineering. Dearborn, Michigan: American Supplier Institute; 1987

[35] Statistics Solutions. ANOVA [WWW Document]. 2013. Retrieved from http://www.statisticssolutions.com/academic-solutions/resources/directory-of-statistical-analyses/anova/

[36] Daneshvar N, Khataee AR, Rasoulifard MH, Pourhassan M. Biodegradation of dye solution containing Malachite green: Optimization of effective parameters using Taguchi method. Journal of Hazardous Materials. 2007;**143**(1-2):214-219

[37] Du Plessis BJ, De Villiers GH. The application of the Taguchi method in the evaluation of mechanical flotation in waste activated sludge thickening. Resources, Conservation and Recycling. 2007;**50**(2):202-210. DOI: 10.1016/j.resconrec.2006.06.014

[38] Tasirin SM, Kamarudin SK, Ghani JA, Lee KF. Optimization of drying parameters of bird's eye chilli in a fluidized bed dryer. Journal of Food Engineering. 2007;**80**(2):695-700

[39] Wu CH, Chen WS. Injection moulding and injection compression moulding of three-beam grating of DVD pickup lens. Sensors and Actuators A: Physical. 2006;**125**(2):367-375

[40] Villafranca RR, Zúnica L, Zúnica RR. Ds-optimal experimental plans for robust parameter design. Journal of Statistical Planning and Inference. 2007;**137**(4):1488-1495

[41] Elshennawy AK. Quality in the new age and the body of knowledge for quality engineers. Total Quality Management and Business Excellence. 2004;**15**(5-6):603-614

[42] Ng EYK, Ng WK. Parametric study of the biopotential equation for breast tumour identification using ANOVA and Taguchi method. Medical and Biological Engineering and Computing. 2006;**44**(1-2):131-139

[43] Houng JY, Liao JH, Wu JY, Shen SC, Hsu HF. Enhancement of asymmetric bio reduction of ethyl 4-chloro acetoacetate by the design of composition of culture medium and reaction conditions. Process Biochemistry. 2007;**42**(1):1-7

[44] Bernardo FP, Pistikopoulos EN, Saraiva PM. Quality costs and robustness criteria in chemical process design optimization. Computers and Chemical Engineering. 2001;**25**:27-4

[45] Kundu A, Gupta BS, Hashim MA, Redzwan G. Taguchi optimization approach for production of activated carbon from phosphoric acid impregnated palm kernel shell by microwave heating. Journal of Cleaner Production. 2015;**105**:420-427. DOI: 10.1016/j.jclepro.2014.06.093

[46] Zolfaghari G, Sari AE, Anbia M, Younesi H, Amirmahmoodi S, Nazari AG. Taguchi optimization approach for Pb(II) and Hg(II) removal from aqueous solutions using modified mesoporous carbon. Journal of Hazardous Materials. 2011;**192**:1046-1055

[47] Mohan SV, Raghavulu SV, Mohanakrishna G, Srikanth S, Sarma PN. Optimization and evaluation of fermentative hydrogen production and wastewater treatment processes using data enveloping analysis (DEA) and Taguchi design of experimental (DOE) methodology. International journal of hydrogen energy. 2009;**34**:216-226. DOI: 10.1016/j.ijhydene.2008.09.044

[48] Mohan SV, Reddy BP, Sarma PN. Ex-situ slurry phase bioremediation of chrysene contaminated soil with the function of metabolic function: process evaluation by data enveloping analysis (DEA) and Taguchi design of experimental methodology (DOE). Bioresource Technology. 2009;**100**(1):164-72. DOI: 10.1016/j.biortech.2008.06.020

[49] Silva MB, Carneiro LM, Silva JPA, Oliveira IS, Filho HJI, Almeida CRO. An application of the Taguchi method (robust design) to environmental engineering: Evaluating advanced oxidative processes in polyester-resin wastewater treatment. American Journal of Analytical Chemistry. 2014;**5**:828-837. DOI: 10.4236/ajac.2014.513092

[50] Liao HT, Shie JR, Yang YK. Applications of Taguchi and design of experiments methods in optimization of chemical mechanical polishing process parameters. International Journal of Advanced Manufacturing Technology. 2008;**38**:674-682. DOI: 10.1007/s00170-007-1124-7

[51] Mohan SV, Rao NC, Prasad KK, Muralikrishna P, Rao RS, Sarma PN. Anaerobic treatment of complex chemical wastewater in a sequencing batch biofilm reactor: Process optimization and evaluation of factors interaction using the Taguchi dynamic DOE methodology. Biotechnology and Bioengineering. 2005;**90**(6):732-745

[52] Mousavi SM, Yaghmaei S, Jafari A, Vossoughi M, Ghobadi Z. Optimization of ferrous biooxidation rate in a packed bed bioreactor using Taguchi approach. Chemical Engineering and Processing: Process Intensification. 2007;**46**(10):935-940. DOI: 10.1016/j.cep.2007.06.010

[53] Kim KD, Han DN, Kim HT. Optimization of experimental conditions based on the Taguchi robust design for the formation of nano-sized silver particles by chemical reduction method. Chemical Engineering Journal. 2004;**104**(1-3):55-61. DOI: 10.1016/j.cej.2004.08.003

[54] Ho CY, Lin ZC. Analysis and application of grey relation and ANOVA in chemical–mechanical polishing process parameters. International Journal of Advanced Manufacturing Technology. 2003;**21**(1):10-14. DOI: 10.1007/s001700300001

[55] Nair VN. Taguchi's parameter design: A panel discussion. Technometrics. 1992;**31**(2):127-161

[56] Montgomery DC. Design and Analysis of Experiments. Singapore: Wiley. 1991

[57] Fraley S, Oom M, Terrien B, Date JZ. Design of Experiments via Taguchi Methods: Orthogonal Arrays: The Michigan Chemical Process Dynamic and Controls Open Text Book. USA. 2006. 678-698

[58] Parks J. On stochastic optimization: Taguchi methods demystified, its limitations and fallacy clarified. Probabilistic Engineering Mechanics. 2006;**16**(1):87-101

[59] Unitek Miyachi Group. Welding material control. Technical Application Brief. 1999;**2**:1-5

Chapter 10

Utilization of Response Surface Methodology in Optimization of Extraction of Plant Materials

Alev Yüksel Aydar

Additional information is available at the end of the chapter

http://dx.doi.org/10.5772/intechopen.73690

Abstract

Experimental design plays an important role in several areas of science and industry. Experimentation is an application of treatments applied to experimental units and is then part of a scientific method based on the measurement of one or more responses. It is necessary to observe the process and the operation of the system well. For this reason, in order to obtain a final result, an experimenter must plan and design experiments and analyzes the results. One of the most commonly used experimental designs for optimization is the response surface methodology (RSM). Because it allows evaluating the effects of multiple factors and their interactions on one or more response variables it is a useful method. In this section, recent studies have been compiled which aim to extraction of plant material in high yield and quality and determine optimum conditions for this extraction process.

Keywords: design of experiments, olive, phenolic content, yield, RSM, food science

1. Introduction

The response surface methodology (RSM) is a widely used mathematical and statistical method for modeling and analyzing a process in which the response of interest is affected by various variables [1] and the objective of this method is to optimize the response [2]. The parameters that affect the process are called dependent variables, while the responses are called dependent variables [3].

For example, the hardness of a meat is affected by cooking time X_1 and cooking temperature X_2. The meat hardness can be changed under any combination of treatment X_1 and X_2. Therefore, time and temperature can vary continuously. If treatments are from a continuous range of values, response surface methodology is useful for developing, improving, and optimizing the response variable. In this case, the hardness of meat Y is the response variable, and it is a function of time and temperature of cooking. It can be expressed as the dependent variable y is a function of X_1 and X_2.

$$Y = f(X_1) + f(X_2) + e \tag{1}$$

where Y is the response (dependent variable), X_1 and X_2 are independent variables and e is the experimental error.

Response surface is a method based on surface placement. Therefore, the main goals of a RSM study are to understand the topography of the response surface including the local maximum, local, minimum and ridge lines and find the region where the most appropriate response occurs [4].

The RSM investigates an appropriate approximation relationship between input and output variables and identify the optimal operating conditions for a system under study or a region of the factor field that satisfies the operating requirements [5, 6]. Box-Behnken designs (BBD) and central composite design (CCD) are two main experimental designs used in response surface methodology [3]. Central composite rotatable design (CCRD) and face central composite design (FCCD) has also been applied to optimization studies in recent years [7–9].

The experimental data are evaluated to fit a statistical model (Linear, Quadratic, Cubic or 2FI (two factor interaction)). The coefficients of the model are represented by constant term, A, B and C (linear coefficients for independent variables), AB, AC and BC (interactive term coefficient), A^2, B^2 and C^2 (quadratic term coefficient). Correlation coefficient (R^2), adjusted determination coefficient (Adj-R^2) and adequate precision are used to check the model adequacies; the model is adequate when its P value < 0.05, lack of fit P value > 0.05, $R^2 > 0.9$ and Adeq Precision >4. Differences between means can be tested for statistical significance using analysis of variance (ANOVA) [10].

1.1. The basic and theoretical aspects of RSM

The design of experiments (DoE) is the most important aspect of RSM. The DoE aims the selection of most suitable points where the response should be well examined. The mathematical model of the process is mostly related to design of experiments. Thus, the selection of experiment design has a great effect in determining the correctness of the response surface construction. The advantages offered by the RSM can be summarized as determining the interaction between the independent variables, modeling the system mathematically, and saving time and cost by reducing the number of trials [11]. However, the most important disadvantage of the response surface method is that the experimental data are fitted to a polynomial model at the second level. It is not correct to say that all systems with curvature are compatible with a second-order polynomial model. In addition, experimental verification of the estimated values in the model should be done absolutely [3].

In early stage of DoE, screening experiments are performed. If there are many variables have little or more effect on the response, the variables which have large effects on response are identified. Therefore the aim is to determine the design variables that have large effects for further investigation [12].

2. RSM application in optimization of extraction

Using Response Surface Method in the extraction studies has been of interest to many researchers in recent years [10, 13, 14]. The steps that must be followed in order to apply this method correctly are shown in **Figure 1**.

Recent optimization studies using the response surface method in extraction from plant materials are summarized in **Table 1**. Independent and dependent variable numbers and the optimization designs are also demonstrated in the same table.

2.1. Yield

Extraction yield is one of the main properties determining efficiency of olive oil extraction. This parameter indirectly takes into account the oil content held in vegetable water and pomace [15, 16].

Extraction yield is defined as the percentage of the extracted olive oil from the total weight of fruit (g). The extraction yield is calculated using the formula below [10]:

$$\text{Yield} = \frac{\textit{Extracted Oil}(g)}{\textit{Olive Fruit}(g)} \text{ x } 100 \quad (2)$$

Aydar et al. used olive fruits (*Olea europaea L.*) from Edremit cultivar grown in Mut area were harvested in the 2015 crop season with a maturity index of 3.35 to obtain ideal conditions for an ultrasound assisted olive oil extraction. It was aimed an extraction for extra virgin olive oils in low acidity and high yield using the Box-Behnken design to optimize extraction parameters including ultrasound time, ultrasound temperature and malaxation time [10].

In terms of yield, the independent variable (X_2), the quadratic term (X_2^2) and the interactive terms (X_1X_2, X_2X_3) were all significant ($P < 0.05$). The quadratic regression model for AV was as follows:

$$\text{Yield} = 7.48 + 0.9062\,5X_2 + 0.8875\,X_3 - 1.1X_1\,X_2 + 0.4375\,X_2\,X_3 - 1.3525\,X_2^{\,2} \quad (3)$$

The most significant effect on the extraction yield ($P < 0.05$) was the malaxation temperature among all ultrasound extraction variables. Conversely, ultrasound time showed no effect ($P > 0.05$) on the yield [10].

The response surface methodology has been applied to determine the optimization of olive paste heating and how it is affected by the independent process variables including olive

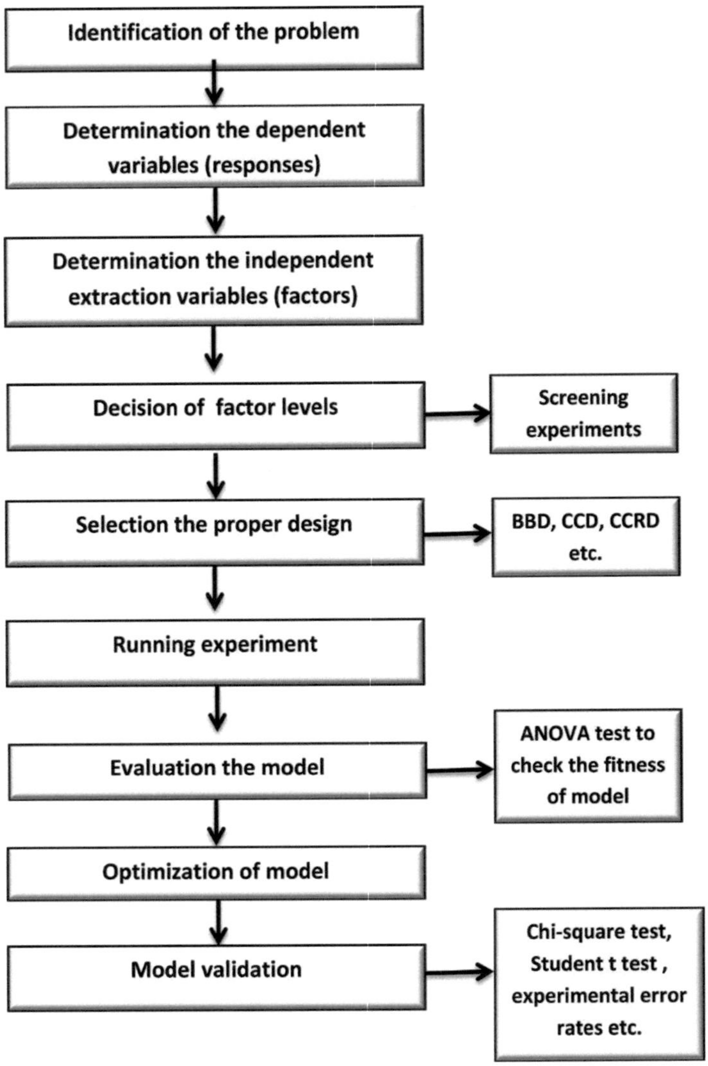

Figure 1. Steps for response surface methodology.

paste flow (Q), high power ultrasound (HPU) intensity (W), olive temperature (OT), olive moisture (OM) and olive fat content (OF) by Bejaoui et al. [17]. They obtained a 2FI (two factor interaction) model for olive paste temperature according to the analysis of variance which showed that the regression model was significant for a P-value <0.0001. The most significant terms of the model were Q, W and the interaction terms Q*W and W*OF based on P-values less than 0.0001 [17].

Second-order equations for oleuropein yield was shown in Eq. (4) [9]

$$\text{Yield} = 0.62767 - 0.029622\,X_1 - 2.60 \times 10^{-3}\,X_2 - 0.056494\,X_3 + 4.26 \times 10^{-5}\,X_1 X_2 + 5.07 \times 10^{-3}\,X_1 X_3 + 2.48 \times 10^{-4}\,X_2 X_3 + 1.15 \times 10^{-4}\,X_{21} + 2.53 \times 10^{-6}\,X_{22} - 0.013423\,X_{23} \quad (4)$$

http://dx.doi.org/10.5772/intechopen.73690

Extraction	Extraction method	Process parameters	Design method	Dependent variables	Model	Ref
Olive leaf	Ultrasound assisted extraction	Solvent concentration, the ratio of solid to solvent, extraction time	BBD	Extract yield, total polyphenol content, antioxidant activity	Quadratic polynomial	[27]
Olive waste	Non-conventional aqueous extraction method	NaOH, temperature, time, mass of the waste	BBD	Total phenolic content, relative color strength	Quadratic polynomial	[29]
Olive leaf	Solvent-free microwave-assisted extraction	Amount of sample, irradiation power, the extraction time.	FCCD	Oleuropein yield and total phenolic content	Quadratic polynomial	[9]
Olive oil	Ultrasound assisted extraction	Ultrasound time, ultrasound temperature, malaxation time	BBD	Oil yield, acidity	Quadratic polynomial	[10]
Olive oil	High power ultrasound assisted extraction	Olive paste flow, ultrasound intensity, fruit temperature before crushing, olive moisture, olive fat content	BBD	Olive paste temperature	2FI	[17]
Olive oil	Conventional extraction	Malaxation time and temperature	CCD	Acidity, peroxide value, K232, K270, Total phenolic content	Quadratic polynomial	[32]
Black Carrot	Ultrasound assisted extraction	Ultrasound energy density, temperature	CCD	Anthocyanin compounds	Quadratic polynomial	[13]
Curry leaf	Ultrasound assisted extraction	Temperature, ultrasonic power, methanol concentration	CCD	Catechin yield, myricetin yield, quercetin yield, antioxidant activity	Quadratic polynomial	[20]
Rapeseed meal	Ultrasound assisted solvent extraction	Temperature, liquid to material ratio, duration and ultrasonic power	BBD	Carotenoid yield	Second-order (Quadratic) polynomial	[31]
Gac fruit peel	Solvent extraction	Extraction time, extraction temperature, solvent ratio	BBD	Total carotenoid, Antioxidant capacity	Quadratic polynomial	[30]
Coffee silverskin	Ultrasound assisted extraction/Microwave assisted extraction	Extraction time, extraction temperature	CCD	Total phenolic content, radical scavenging capacity, total caffeoylquinic acids, caffeine content	Quadratic polynomial	[21]

Extraction	Extraction method	Process parameters	Design method	Dependent variables	Model	Ref
Brown seaweed	Ultrasound assisted extraction	Extraction time, acid concentration, ultrasound amplitude	BBD	Total phenolic, fucose, uronic acids	Second-order (Quadratic) polynomial	[35]
Hazelnut skin	Ultrasound assisted extraction	Extraction time, temperature, ultrasound amplitude	CCD, BBD	Crude polysaccharide yield, consumed energy	Quadratic polynomial	[25]
Trapa quadrispinosa stems	Ultrasound assisted extraction	Ultrasonic time, liquid to material ratio, ultrasonic temperature	BBD	Polysaccharide yield, Ferric-Reducing Antioxidant Capacity (FRAC)	Quadratic polynomial	[26]
Sphallerocarpus gracilis roots	Hot water extraction, Ultrasound assisted extraction	Extraction temperature, Extraction time, Liquid–solid ratio, Ultrasound power	BBD	*S. gracilis* yield	Quadratic polynomial	[33]
Papaya seed oil	Ultrasound assisted extraction	Time, temperature, ultrasound power, solvent to sample ratio	SCCD	Yield, antioxidant activity, p-anisidine value, peroxide value, totox value	Quadratic polynomial	[18]
Pomegranate seed oil	Ultrasound assisted extraction	Ultrasonic power, extraction temperature, extraction time, the ratio of solvent volume and seed weight	BBD	Oil yield	Quadratic polynomial	[19]

Table 1. Summary of recent studies published on the extraction of plant materials optimized by RSM.

Where X_1 is the amount of sample, X_2 is the Microwave (MW) irradiation power, and X_3 is the extraction time. The researchers found that the second power of microwave intensity was the most significant parameter, followed by the amount of sample, quadratic time, and power for oleuropein yield [9].

Response surface method has been used frequently in recent years to optimize different oil extractions other than olive oil including papaya seed oil and pomegranate seed oil [18, 19].

To optimize the ultrasound-assisted extraction conditions followed by ultrahigh performance liquid chromatography (UHPLC) to achieve high catechin, myricetin, and quercetin contents, and high antioxidant and anticancer activities in the curry leaf extracts, RSM was applied by Ghasemzadeh et al. [20]. They used the central composite experimental design (3-level, 3-factorial) to determine the optimum extraction parameters affecting the extraction yields of catechin (Y1), myricetin (Y2), quercetin (Y3), and antioxidant activity (Y4) of curry leaf extracts [20].

The extraction efficiency of UAE and MAE methods was compared to a conventional solvent extraction by Guglielmetti et al. [21]. Authors used RSM with a CCD to investigate ultrasound assisted extraction (UAE) and microwave assisted extraction (MAE) of caffeoylquinic acids and caffeine from coffee silverskin (CS) at two particle size. They found that the highest caffeine content (14.24 g kg^{-1} dw) with a significant reduction of extraction time was obtained by UAE [21].

Since different extraction methods have important impacts on the polysaccharide bioactivity, yield and structure, to find the best extraction method to obtain high yield of polysaccharide is crucial. Recently several researchers used RSM for optimization of polysaccharide extraction from different plant materials [22–26]. To investigate the best response surface design for optimization of polysaccharide yield (CPS) from hazelnut skin, CCD and BBD designs were studied by Yılmaz and Tavman [25]. Optimum conditions for a maximum yield of polysaccharide extraction from *Trapa quadrispinosa* stems recently determined by Raza et al. 41 min, 31.5 mL/g and 58°C were the optimum conditions for extraction time, ratio of water to material, and extraction temperature, respectively [26].

2.2. Phenolic and antioxidant compound extraction from plant materials

In recent years, there has been a growing interest in finding new natural sources of food antioxidants. As a main fruit crop, olive is also valued due to its phenolic- containing leaves. Optimization of ultrasound-assisted extraction of olive leaf has been studied by extraction parameters including solid/solvent ratio, time and ethanol concentration by Şahin and Şamli [27]. In order to obtain the maximum extraction performance for an ultrasound assisted extraction, 500 mg olive leaf to 10 mL solvent ratio, 60 min of extraction time and 50% ethanol composition were found to be as optimal operating conditions [27].

Shirzad et al. also studied on optimization of olive leave extraction in order to shorten the time of extraction and decrease the consumption of energy. The conditions for obtaining maximum yield of polyphenols, total flavonoids and antioxidants were optimized using RSM The effects of ultrasonic temperature (35–65°C), ultrasonic time (5–15 min), and ethanol to water ratio (Et:W) (25–75%) were evaluated. The highest extraction yield was found to be 51% of ethanol to water ratio at 65°C for 15 min [28].

Elksibi et al. used RSM to investigate the optimization of natural colorant non-conventional extraction technique from olive waste. They studied the combined effects of extraction conditions on total phenolic content (TPC) and relative color strength (K/S) using a three-level three-factor Box-Behnken design [29].

Second-order equation for total phenolic content from olive leaf obtained by RSM was shown in Eq. (5) by Şahin et al. [9]:

$$\begin{aligned} \text{TPC} = & -0.019369 - 0.3600\,3X_1 + 0.1424\,9X_2{}^{-13}.6102\,9X_3 + 6.64 \times 10^{-4} X_1 X_2 + 0.089174\, X_1 X_3 \\ & + 4.53 \times 10^{-3} X_2 X_3 - 0.012889\, X_{21} - 2.74 \times 10^{-4} X_{22} + 2.3499\,3X_{23} \end{aligned} \quad (5)$$

where X_1 is the amount of sample, X_2 is the MW irradiation power, and X_3 is the extraction time [9].

Agcam et al. [13] used response surface methodology to optimize ultrasound assisted anthocyanin compounds extraction from black carrot. The independent variables were temperature and ultrasound energy density which is calculated with following Eq. (6):

$$E = \frac{P.t}{M} \quad (6)$$

The optimization of five different anthocyanin compounds from black carrot was conducted using CCD design with a 16 factorial experiments, 5 replicates of the central point. They obtained quadratic polynomial equations for each anthocyanin compound which were cyanidin-3-xylosyl-glucosyl-galactoside (C3XGG),cyanidin-3-xylosyl-galactoside (C3XG), monoacylated anthocyanins cyaniding-3-xylosyl-glucosyl-galactosidesinapic acid (C3XGGS), cyanidin-3-xylosyl-glucosylgalactoside-ferulic acid (C3XGGF), and cyanidin-3-xylosyl-glucosyl-galactoside-coumaric acid (C3XGGC) [13].

Ghasemzadeh et al. [20] found that ANOVA for predicted model of antioxidant activity was significant (F-value 17.21, $P < 0.0001$) with a good coefficient of determination ($R^2 = 0.98$). They also observed that extraction variables showed significant ($P < 0.01$) quadratic and linear effects on the antioxidant activity and the predicted model obtained for DPPH (Y_4) was as follows:

$$\begin{aligned} \text{DPPH} = & +79.56 - 5.70\, X_1 + 1.88\, X_2 + 1.29\, X_3 - 1.31\, X_1 X_2 + 0.24\, X_1 X_3 + 0.64\, X_2 X_3 \\ & - 15.29\, X_1{}^2 - 0.57\, X_2{}^2 - 1.14\, X_3{}^2 \end{aligned} \quad (7)$$

Where X_1 is the temperature, X_2 is the methanol concentration, and X_3 is the ultrasonic power.

Using RSM the extraction conditions including extraction time, temperature and solvent–solid ratio were optimized for maximizing extraction yields of carotenoids and antioxidant capacity from Gac fruit peel by Chuyen et al. [30]. In that study most effective solvent was ethyl acetate and optimal extraction conditions (time, temperature and solvent-solid ratio) were 150 min, 40.7°C and 80 mL g^{-1}, respectively [30].

Box-Behnken design (BBD) with a total number of 29 experiments were conducted for four factors (temperature, liquid to material ratio, duration and ultrasonic power) and at three levels to obtain high yield of carotenoid from rapeseed meal. Optimal ultrasound assisted extraction conditions were as follows: temperature 49.6°C, liquid to material ratio 41.4 mL/g, duration 48.5 min, ultrasonic power 252.9 W [31].

Guglielmetti et al. observed a positive correlation between an increase of temperature and total phenolic content (TPC) for conventional solvent extraction and UAE; a negative effect on TPC when using MAE above 50°C. They found that temperature was the most effective process variable on extraction processes [21].

Espínola et al. used RSM to investigate the optimum extraction condition for virgin olive oil extraction from olives at three different maturation index (MI). In olives at lowest maturity index, temperature had a positive effect on polyphenol content at low malaxation temperatures, however no significant effect was determined at higher temperatures. On the contrary, malaxation time had a slight influence at lower temperatures. In higher MI olives, variations of polyphenol content were not significantly different [32].

3. Validation of the model

In the response surface method, the model that best represents how dependent variables are affected by independent variables is determined theoretically. However, experiments should be carried out to verify the reliability of the theoretically determined models under optimum conditions. Chi-Square test and t-tests are most commonly used to determine the difference between experimental and predicted values. Another method to evaluate the validation of model is to calculate experimental error between theoretical and experimental values.

The experimental and predicted values were 8.31 and 8.42% for the acidity and the yield were 0.31 g oleic acid/ 100 g olive oil and 0.28 g oleic acid/ 100 g olive oil for predicted and experimental values, respectively. These results were in good agreement with the predicted values under the optimum working condition. Therefore, the acidity value of olive oil and yield for any combination of ultrasound time, ultrasound temperature and malaxation time could be accurately predictor by the regression models obtained by RSM [10]. In the 2005–2006 season, the estimated extraction yield, acidity and peroxide index of the 3.2 MI olive samples showed that the experimental data were consistent with the model for all three dependent variables [32].

Elksibi et al. found that experimental value of 22.54 and 1120 mg/L for the color strength parameter (K/S) and the total phenolic content, respectively. While the predicted values were 23.22 and 1134 mg/L for the color strength parameter (K/S) and the total phenolic content, respectively. They determined the results obtained at the optimal combination was in agreement with the theoretical result. Therefore, the model obtained in this research was confirmed [29].

The experimental extraction yield in the hot water extraction process was 3.79 ± 0.13% and the yield in the ultrasound extraction process was 6.04 ± 0.21% under the optimum conditions, which were in good agreement with the predicted values. These results demonstrated that the extraction models were reliable and accurate [33].

15 min, 45°C and 50% amplitude was selected as an optimal level of parameters to validate the result of desirability functions. 1.69% CPS yield and 73.00 kJ energy consumption were found and the predicted values obtained by CCD and BBD were similar to the experimental values and the points of all predicted and experimental response values were correlating. Thus the model developed was significant and reliable. Studentized test results were in agreement with experimental runs which showed that all the data points were kept within the limits [34].

Validation of the regression equation and statistical model was conducted at 49.6°C, 41.4 mL/g, 48.5 min and 240 W which were temperature, liquid to material ratio, extraction time and power of ultrasound, respectively. With these optimized conditions, the predicted response for carotenoid yield was approximately 0.1570 mg/g, and the experimental value was found as 0.1577 ± 0.0014 mg/g. These results confirmed that experimental values are in agreement with the predicted values, thus the model was validated [31].

4. Conclusions

Response surface methodology with a wide range of applications in food science and technology has been successfully used for many years. Optimization of the extraction of plant materials known to be useful for health has attracted many researchers in recent years. This section summarizes the recent researches that optimize extraction conditions necessary to obtain higher quality and yield than plant materials using RSM. One of the most important points in the implementation of this method is that the predicted values in the model should be verified experimentally. RSM has many advantages when compared to classical methods. It needs fewer experiments to study the effects of all the factors and the optimum combination of all the variables can be revealed. The interaction (the behavior of one factor may be dependent on the level of another factor) between factors can be determined. It also requires less time and effort. With all of these advantages, it will be used not only in food science but also in other areas in future.

Author details

Alev Yüksel Aydar

Address all correspondence to: alevyuksel.aydar@cbu.edu.tr

Department of Food Engineering, Faculty of Engineering, Manisa Celal Bayar University, Muradiye, Manisa, Turkey

References

[1] Refinery NP, Braimah MN. Utilization of response surface methodology (RSM) in the optimization of crude oil refinery. Journal of Multidisciplinary Engineering Science and Technology (JMEST). 2016;**3**:4361-4369

[2] Montgomery DC. Design and Analysis of Experiments: Response Surface Method and Designs. New Jersey: John Wiley and Sons, Inc; 2005

[3] Koç B, Kaymak-Ertekin F. Response surface methodology and food processing applications. Gıda. 2009;**7**:1-8

[4] Bradley N. The response surface methodology. (M.A.) thesis. Indiana University South Bend; 2007

[5] Farooq Z, Rehman S, Abid M. Application of response surface methodology to optimize composite flour for the production and enhanced storability of leavened flat bread (Naan). Journal of Food Processing and Preservation. 2013;**37**:939-945

[6] Pishgar-Komleh SH, Keyhani A, Msm R, Jafari A. Application of response surface methodology for optimization of Picker-Husker harvesting losses in corn seed. Iranica Journal of Energy and Environment. 2012;**3**(2):134-142

[7] Wang J, Sun B, Cao Y, Tian Y, Li X. Optimisation of ultrasound-assisted extraction of phenolic compounds from wheat bran. Food Chemistry. 2008;**106**:804-810

[8] Prakash Maran J, Mekala V, Manikandan S. Modeling and optimization of ultrasound-assisted extraction of polysaccharide from *Cucurbita moschata*. Carbohydrate Polymers. 2013;**92**:2018-2026

[9] Şahin S, Samli R, Tan AS, Barba FJ, Chemat F, Cravotto G, Lorenzo JM. Solvent-free microwave-assisted extraction of polyphenols from olive tree leaves: Antioxidant and antimicrobial properties. Molecules. 2017;**7**:1054

[10] Aydar AY, Bağdatlıoğlu N, Köseoğlu O. Effect of ultrasound on olive oil extraction and optimization of ultrasound-assisted extraction of extra virgin olive oil by response surface methodology (RSM). Grasas Y Aceites = International journal of fats and oils. 2017;**68**

[11] Boyacı İH. A new approach for determination of enzyme kinetic constants using response surface methodology. Biochemical Engineering Journal. 2005;**25**:55-62

[12] Nagendra S. Response surface methodology: Advantages and challenges. Design Optimization. Schenectady, New York, USA: General Electric Corporate R&D; pp. 100-104. https://www.brad.ac.uk/staff/vtoropov/DO/vol1/Forum_i1_100-104.pdf

[13] Agcam E, Akyıldız A, Balasubramaniam VM. Optimization of anthocyanins extraction from black carrot pomace with thermosonication. Food Chemistry. 2017;**237**:461-470

[14] Chaiklahan R, Chirasuwan N, Triratana P, Loha V, Tia S, Bunnag B. Polysaccharide extraction from Spirulina sp. and its antioxidant capacity. International Journal of Biological Macromolecules. 2013;**58**:73-78

[15] Angerosa F, Mostallino R, Basti C, Vito R. Influence of malaxation temperature and time on the quality of virgin olive oils. Food Chemistry. 2001;**72**:19-28

[16] Puértolas E, Martínez de Marañón I. Olive oil pilot-production assisted by pulsed electric field: Impact on extraction yield, chemical parameters and sensory properties. Food Chemistry. 2015;**167**:497-502

[17] Bejaoui MA, Beltran G, Aguilera MP, Jimenez A. Continuous conditioning of olive paste by high power ultrasounds: Response surface methodology to predict temperature and its effect on oil yield and virgin olive oil characteristics. LWT- Food Science and Technology. 2016;**69**:175-184

[18] Samaram S, Mirhosseini H, Tan CP, Ghazali HM, Bordbar S, Serjouie A. Optimisation of ultrasound-assisted extraction of oil from papaya seed by response surface methodology: Oil recovery, radical scavenging antioxidant activity, and oxidation stability. Food Chemistry. 2015;**172**:7-17

[19] Tian Y, Xu Z, Zheng B, Martin Lo Y. Optimization of ultrasonic-assisted extraction of pomegranate (*Punica granatum* L.) seed oil. Ultrasonics Sonochemistry. 2013;**20**:202-208

[20] Ghasemzadeh A, Jaafar HZ, Karimi E, Rahmat A. Optimization of ultrasound-assisted extraction of flavonoid compounds and their pharmaceutical activity from curry leaf (*Murraya koenigii* L.) using response surface methodology. BMC Complementary and Alternative Medicine. 2014;**14**:318

[21] Guglielmetti A, Ghirardello D, Belviso S, Zeppa G. Optimisation of ultrasound and microwave-assisted extraction of caffeoylquinic acids and caffeine from coffee silverskin using response. Italian Journal of Food Science. 2017;**29**:409-423

[22] Mei L, Zhen-Chang W, Hao-Jie D, Li C, Qing-Gang X, Jing L. Response surface optimization of polysaccharides extraction from Liriope roots and its modulatory effect on Sjogren syndrome. International Journal of Biological Macromolecules. 2009;**45**:284-288

[23] Chen R, Li Y, Dong H, Liu Z, Li S, Yang S, Li X. Optimization of ultrasonic extraction process of polysaccharides from Ornithogalum Caudatum Ait and evaluation of its biological activities. Ultrasonics Sonochemistry. 2012;**19**:1160-1168

[24] Zhu C, Zhai X, Li L, Wu X, Li B. Response surface optimization of ultrasound-assisted polysaccharides extraction from pomegranate peel. Food Chemistry. 2015;**177**:139-146

[25] Yilmaz T, Tavman Ş. Ultrasound assisted extraction of polysaccharides from hazelnut skin. Food Science and Technology International. 2015;**22**(2):112-121

[26] Raza A, Li F, Xu X, Tang J. Optimization of ultrasonic-assisted extraction of antioxidant polysaccharides from the stem of *Trapa quadrispinosa* using response surface methodology. International Journal of Biological Macromolecules. 2017;**94**:335-344

[27] Şahin S, Şamli R. Optimization of olive leaf extract obtained by ultrasound-assisted extraction with response surface methodology. Ultrasonics Sonochemistry. 2013;**20**:595-602

[28] Shirzad H, Niknam V, Taheri M, Ebrahimzadeh H. Ultrasound-assisted extraction process of phenolic antioxidants from olive leaves: A nutraceutical study using RSM and LC–ESI–DAD–MS. Journal of Food Science and Technology. 2017;**54**:2361-2371

[29] Elksibi I, Haddar W, Ticha MB, Mhenni MF. Development and optimisation of a non conventional extraction process of natural dye from olive solid waste using response surface methodology (RSM). Food Chemistry. 2016;**161**:345-352

[30] Chuyen HV, Roach PD, Golding JB, Parks SE, Nguyen MH. Optimisation of extraction conditions for recovering carotenoids and antioxidant capacity from Gac peel using response surface methodology. International Journal of Food Science and Technology. 2017;**52**:972-980

[31] Yan F, Fan K, He J, Gao M. Ultrasonic-assisted solvent extraction of carotenoids from rapeseed meal: Optimization using response surface methodology. Journal of Food Quality. 2015;**38**:377-386

[32] Espínola F, Moya M, Fernández DG, Castro E. Modelling of virgin olive oil extraction using response surface methodology. International Journal of Food Science and Technology. 2011;**46**:2576-2583

[33] Ma T, Sun X, Tian C, Luo J, Zheng C, Zhan J. Polysaccharide extraction from *Sphallerocarpus gracilis* roots by response surface methodology. International Journal of Biological Macromolecules. 2016;**88**:162-170

[34] Yılmaz T, Tavman S. Modeling and optimization of ultrasound assisted extraction parameters using response surface methodology for water soluble polysaccharide extraction from hazelnut skin. Journal of Food Processing & Preservation. 2016;**41**:1-13

[35] Kadam SU, Tiwari BK, Smyth TJ, O'Donnell CP. Optimization of ultrasound assisted extraction of bioactive components from brown seaweed Ascophyllum nodosum using response surface methodology. Ultrasonics Sonochemistry. 2015;**23**:308-316